HIGHWAY ENGINEERING

HIGHWAY ENGINEERING

Martin Rogers
Department of Civil and Structural Engineering
Dublin Institute of Technology
Ireland

Blackwell
Science

© 2003 by Blackwell Publishing Ltd
Editorial Offices:
9600 Garsington Road, Oxford OX4 2DQ
   Tel: +44 (0) 1865 776868
108 Cowley Road, Oxford OX4 1JF, UK
   Tel: +44 (0)1865 791100
Blackwell Publishing USA, 350 Main
Street, Malden, MA 02148-5018, USA
   Tel: +1 781 388 8250
Iowa State Press, a Blackwell Publishing
Company, 2121 State Avenue, Ames, Iowa
50014-8300, USA
   Tel: +1 515 292 0140
Blackwell Munksgaard, 1 Rosenørns Allé,
P.O. Box 227, DK-1502 Copenhagen V,
Denmark
   Tel: +45 77 33 33 33
Blackwell Publishing Asia Pty Ltd,
550 Swanston Street, Carlton South,
Victoria 3053, Australia
   Tel: +61 (0)3 9347 0300
Blackwell Verlag, Kurfürstendamm 57,
10707 Berlin, Germany
   Tel: +49 (0)30 32 79 060
Blackwell Publishing, 10 rue Casimir
Delavigne, 75006 Paris, France
   Tel: +33 1 53 10 33 10

First published 2003

A catalogue record for this title is available from the
British Library

ISBN 0-632-05993-1

Library of Congress
Cataloging-in-Publication Data
Rogers, Martin.
   Highway engineering / Martin Rogers. – 1st ed.
      p. cm.
   ISBN 0-632-05993-1 (Paperback : alk. paper)
   1. Highway engineering.   I. Title.
   TE145.R65 2003
   625.7 – dc21
                                        2003005910

Set in 10 on 13 pt Times
by SNP Best-set Typesetter Ltd., Hong Kong
Printed and bound in Great Britain by
TJ International Ltd, Padstow, Cornwall

For further information on
Blackwell Publishing, visit our website:
www.blackwellpublishing.com

To Margaret, for all her love, support and encouragement

Contents

Preface

Given the problems of congestion in built-up urban areas, maximising the efficiency with which highways are planned, analysed, designed and maintained is of particular concern to civil engineering practitioners and theoreticians. This book is designed as an introductory text which will deliver basic information in those core areas of highway engineering of central importance to practising highway engineers.

Highway Engineering is intended as a text for undergraduate students on degree and diploma courses in civil engineering. It does, however, touch on topics which may be of interest to surveyors and transport planners. The book does not see itself as a substitute for courses in these subject areas, rather it demonstrates their relevance to highway engineering.

The book must be focused on its primary readership – first and foremost it must provide an essential text for those wishing to work in the area, covering all the necessary basic foundation material needed for practitioners at the entry level to industry. In order to maximise its effectiveness, however, it must also address the requirements of additional categories of student: those wishing to familiarise themselves with the area but intending to pursue another speciality after graduation and graduate students requiring necessary theoretical detail in certain crucial areas.

The aim of the text is to cover the basic theory and practice in sufficient depth to promote basic understanding while also ensuring as wide a coverage as possible of all topics deemed essential to students and trainee practitioners. The text seeks to place the topic in context by introducing the economic, political, social and administrative dimensions of the subject. In line with its main task, it covers central topics such as geometric, junction and pavement design while ensuring an adequate grasp of theoretical concepts such as traffic analysis and economic appraisal.

The book pays frequent reference to the Department of Transport's *Design Manual for Roads and Bridges* and moves in a logical sequence from the planning and economic justification for a highway, through the geometric design and traffic analysis of highway links and intersections, to the design and maintenance of both flexible and rigid pavements. To date, texts have concentrated on either highway planning/analysis or on the pavement design and maintenance aspects of highway engineering. As a result, they tend to be advanced in nature rather than introductory texts for the student entering the field of study for the first time. This text aims to be the first UK textbook that meaningfully addresses both traffic planning/analysis and pavement design/maintenance areas within one basic introductory format. It can thus form a platform from which the student can move into more detailed treatments of the different areas of highway engineering dealt with more comprehensively within the more focused textbooks.

Chapter 1 defines highway planning and details the different forms of decision frameworks utilised within this preparatory process, along with the importance of public participation. Chapter 2 explains the basic concepts at the basis of traffic demand modelling and outlines the four-stage transport modelling process. Chapter 3 details the main appraisal procedures, both monetary and non-monetary, required to be implemented in order to assess a highway proposal. Chapter 4 introduces the basic concepts of traffic analysis and outlines how the capacity of a highway link can be determined. Chapter 5 covers the analysis of flows and capacities at the three major types of intersection – priority intersections, signalised junctions and roundabouts. The concepts of design speed, sight distances, geometric alignment (horizontal and vertical) and geometric design are addressed in Chapter 6. Chapter 7 deals with highway pavement materials and the design of both rigid and flexible pavements, while Chapter 8 explains the basics of structural design for highway pavement thicknesses. Finally, the concluding chapter (Chapter 9) takes in the highway maintenance and overlay design methods required as the pavement nears the end of its useful life.

In overall terms, the text sets out procedures and techniques needed for the planning, design and construction of a highway installation, while setting them in their economic and political context.

Every effort has been made to ensure the inclusion of information from the most up-to-date sources possible, particularly with reference to the most recent updates of the *Design Manual for Roads and Bridges*. However, the regularity with which amendments are introduced is such that, by the time this text reaches the bookshelves, certain aspects may have been changed. It is hoped, however, that the basic approaches underlying the text will be seen to remain fully valid and relevant.

The book started life as a set of course notes for a highways module in the civil degree programme in the Dublin Institute of Technology, heavily influenced by my years in practice in the areas of highway planning, design and construction. I am indebted to my colleagues John Turner, Joe Kindregan, Ross Galbraith, Liam McCarton and Bob Mahony for their help and encouragement. My particular gratitude is expressed to Margaret Rogers, partner and fellow professional engineer, for her patience and support. Without her, this book would never have come to exist.

Martin Rogers
Dublin Institute of Technology

Acknowledgements

Extracts from British Standards are reproduced with the permission of the British Standards Institution. BSI publications can be obtained from BSI Customer Services, 389 Chiswick High Road, London W4 4AL, United Kingdom. Tel. +44 (0) 20 8996 9001. Email: cservices@bsi-global.com

Extracts from Special Report 209 of the Highway Capacity Manual (1985) are reproduced with permission of the Transportation Research Board, National Research Council, Washington, DC.

Crown copyright material is reproduced with the permission of the Controller of HMSO and the Queen's Printer for Scotland.

Chapter 1

The Transportation Planning Process

1.1 Why are highways so important?

Highways are vitally important to a country's economic development. The construction of a high quality road network directly increases a nation's economic output by reducing journey times and costs, making a region more attractive economically. The actual construction process will have the added effect of stimulating the construction market.

1.2 The administration of highway schemes

The administration of highway projects differs from one country to another, depending upon social, political and economic factors. The design, construction and maintenance of major national primary routes such as motorways or dual carriageways are generally the responsibility of a designated government department or an agency of it, with funding, in the main, coming from central government. Those of secondary importance, feeding into the national routes, together with local roads, tend to be the responsibility of local authorities. Central government or an agency of it will usually take responsibility for the development of national standards.

The Highways Agency is an executive organisation charged within England with responsibility for the maintenance and improvement of the motorway/trunk road network. (In Ireland, the National Roads Authority has a similar function.) It operates on behalf of the relevant government minister who still retains responsibility for overall policy, determines the framework within which the Agency is permitted to operate and establishes its goals and objectives and the time frame within which these should take place.

In the United States, the US Federal Highways Agency has responsibility at federal level for formulating national transportation policy and for funding major projects that are subsequently constructed, operated and maintained at state level. It is one of nine primary organisational units within the US Department of Transportation (USDOT). The Secretary of Transportation, a member of the President's cabinet, is the USDOT's principal.

Each state government has a department of transportation that occupies a pivotal position in the development of road projects. Each has responsibility for the planning, design, construction, maintenance and operation of its federally funded highway system. In most states, its highway agency has responsibility for developing routes within the state-designated system. These involve roads of both primary and secondary state-wide importance. The state department also allocates funds to local government. At city/county level, the local government in question sets design standards for local roadways as well as having responsibility for maintaining and operating them.

1.3   Sources of funding

Obtaining adequate sources of funding for highways projects has been an ongoing problem throughout the world. Highway construction has been funded in the main by public monies. However, increasing competition for government funds from the health and education sector has led to an increasing desire to remove the financing of major highway projects from competition for government funds by the introduction of user or toll charges.

Within the United Kingdom, the New Roads and Streetworks Act 1991 gave the Secretary of State for Transport the power to create highways using private funds, where access to the facility is limited to those who have paid a toll charge. In most cases, however, the private sector has been unwilling to take on substantial responsibility for expanding the road network within the UK. Roads tend still to be financed from the public purse, with central government fully responsible for the capital funding of major trunk road schemes. For roads of lesser importance, each local authority receives a block grant from central government that can be utilised to support a maintenance programme at local level or to aid in the financing of a capital works programme. These funds will supplement monies raised by the authority through local taxation. A local authority is also permitted to borrow money for highway projects, but only with central government's approval.

Within the US, fuel taxes have financed a significant proportion of the highway system, with road tolls being charged for use of some of the more expensive highway facilities. Tolling declined between 1960 and 1990, partly because of the introduction of the Interstate and Defense Highway Act in 1956 which prohibited the charging of tolls on newly constructed sections of the interstate highways system, but also because of the wide availability of federal funding at the time for such projects. Within the last ten years, however, use of toll charges as a method of highway funding has returned.

The question of whether public or private funding should be used to construct a highway facility is a complex political issue. Some feel that public ownership of all infrastructure is a central role of government, and under no circumstances should it be constructed and operated by private interests. Others take the view that any measure which reduces taxes and encourages private enterprise should be encouraged. Both arguments have some validity, and any responsible government must strive to strike the appropriate balance between these two distinct forms of infrastructure funding.

Within the UK, the concept of design-build-finance-operate (DBFO) is gaining credence for large-scale infrastructure projects formerly financed by government. Within this arrangement, the developer is responsible for formulating the scheme, raising the finance, constructing the facility and then operating it for its entire useful life. Such a package is well suited to a highway project where the imposition of tolls provides a clear revenue-raising opportunity during its period of operation. Such revenue will generate a return on the developer's original investment.

Increasingly, highway projects utilising this procedure do so within the Private Finance Initiative (PFI) framework. Within the UK, PFI can involve the developer undertaking to share with the government the risk associated with the proposal before approval is given. From the government's perspective, unless the developer is willing to take on most of this risk, the PFI format may be inappropriate and normal procedures for the awarding of major infrastructure projects may be adopted.

1.4   Highway planning

1.4.1   Introduction

The process of transportation planning entails developing a transportation plan for an urban region. It is an ongoing process that seeks to address the transport needs of the inhabitants of the area, and with the aid of a process of consultation with all relevant groups, strives to identify and implement an appropriate plan to meet these needs.

The process takes place at a number of levels. At an administrative/political level, a transportation policy is formulated and politicians must decide on the general location of the transport corridors/networks to be prioritised for development, on the level of funding to be allocated to the different schemes and on the mode or modes of transport to be used within them.

Below this level, professional planners and engineers undertake a process to define in some detail the corridors/networks that comprise each of the given systems selected for development at the higher political level. This is the level at which what is commonly termed a 'transportation study' takes place. It defines the links and networks and involves forecasting future population and economic growth, predicting the level of potential movement within the area and describing both the physical nature and modal mix of the system required to cope with the region's transport needs, be they road, rail, cycling or pedestrian-based. The methodologies for estimating the distribution of traffic over a transport network are detailed in Chapter 2.

At the lowest planning level, each project within a given system is defined in detail in terms of its physical extent and layout. In the case of road schemes, these functions are the remit of the design engineer, usually employed by the roads authority within which the project is located. This area of highway engineering is addressed in Chapters 4 to 7.

The remainder of this chapter concentrates on systems planning process, in particular the travel data required to initiate the process, the future planning strategy assumed for the region which will dictate the nature and extent of the network derived, a general outline of the content of the transportation study itself and a description of the decision procedure which guides the transport planners through the systems process.

1.4.2   Travel data

The planning process commences with the collection of historical traffic data covering the geographical area of interest. Growth levels in past years act as a strong indicator regarding the volumes one can expect over the chosen future time, be it 15, 20 or 30 years. If these figures indicate the need for new/upgraded transportation facilities, the process then begins of considering what type of transportation scheme or suite of schemes is most appropriate, together with the scale and location of the scheme or group of schemes in question.

The demand for highway schemes stems from the requirements of people to travel from one location to another in order to perform the activities that make up their everyday lives. The level of this demand for travel depends on a number of factors:

- The location of people's work, shopping and leisure facilities relative to their homes
- The type of transport available to those making the journey
- The demographic and socio-economic characteristics of the population in question.

Characteristics such as population size and structure, number of cars owned per household and income of the main economic earner within each household tend to be the demographic/socio-economic characteristics having the most direct effect on traffic demand. These act together in a complex manner to influence the demand for highway space.

As an example of the relationship between these characteristics and the change in traffic demand, let us examine Dublin City's measured growth in peak travel demand over the past ten years together with the levels predicted for the next ten, using figures supplied by the Dublin Transport Office (DTO) in 2000. Table 1.1 shows that between 1991 and 1999 peak hour demand grew by 65%.

Demand for travel	1991	1999	2016
Thousand person trips (morning peak hour)	172	283	488

Table 1.1 Increase in travel demand within Dublin City, 1991–2016

It has been predicted by DTO that between 1999 and 2016 a further 72.4% of growth will take place.

The cause of these substantial increases can be seen when one examines the main factors influencing traffic growth – population, number of cars per household and economic growth. Between 1991 and 1999, the population within the area increased by just over 8%, and car ownership by 38.5%, with gross domestic product increasing to 179% of its 1991 value. DTO have predicted that, between 1999 and 2016, population will increase by 20% and car ownership by 40%, with gross domestic product increasing to 260% of its 1991 value (see Table 1.2).

Table 1.2 Factors influencing traffic growth within Dublin City, 1991–2016

	1991	1999	2016
Population (million)	1.35	1.46	1.75
Car ownership (per 1000 population)	247	342	480
% Growth in gross domestic product since 1991	—	79%	260%

The significant growth indicated in Table 1.2 is consistent with the past recorded and future predicted traffic demand figures given in Table 1.1. High levels of residential and employment growth will inevitably result in increased traffic demand as more people link up to greater employment opportunities, with the higher levels of prosperity being reflected in higher levels of car ownership. Increasing numbers of jobs, homes, shopping facilities and schools will inevitably increase the demand for traffic movement both within and between centres of population.

On the assumption that a road scheme is selected to cater for this increased future demand, the design process requires that the traffic volumes for some year in the future, termed the design year, can be estimated. (The design year is generally taken as 10–15 years after the highway has commenced operation.) The basic building block of this process is the *current level of traffic* using the section of highway at present. Onto this figure must be added an estimate for the *normal traffic growth*, i.e. that which is due to the year-on-year annual increases in the number of vehicles using the highway between now and the design year. Table 1.1 shows the increase in vehicle trips predicted within the Dublin Region for the first 16 years of the new millennium. Onto these two constituents of traffic volume must be added *generated traffic* – those extra trips brought about directly from the construction of the new road. Computation of these three components enables the design-year volume of traffic to be estimated for the proposed highway. Within the design process, the design volume will determine directly the width of the travelled pavement required to deal with the estimated traffic levels efficiently and effectively.

1.4.3   Highway planning strategies

When the highway planning process takes place within a large urban area and other transport options such as rail and cycling may be under consideration alongside car-based ones, the procedure can become quite complex and the workload involved in data collection can become immense. In such circumstances, before a comprehensive study can be undertaken, one of a number of broad strategy options must be chosen:

● The land use transportation approach
● The demand management approach
● The car-centred approach
● The public transport-centred approach.

Land use transportation approach

Within this method, the management of land use planning is seen as the solution to controlling the demand for transport. The growing trend where many commuters live in suburbs of a major conurbation or in small satellite towns while working within or near the city centre has resulted in many using their private car for their journey to work. This has led to congestion on the roads and the need for both increased road space and the introduction of major public transport improvements. Land use strategies such as the location of employment opportunities close to large residential areas and actively limiting urban sprawl which tends to increase the dependency of commuters on the private car, are all viable land use control mechanisms.

The demand management approach

The demand management approach entails planning for the future by managing demand more effectively on the existing road network rather than constructing new road links. Demand management measures include the tolling of heavily trafficked sections of highway, possibly at peak times only, and car pooling, where high occupancy rates within the cars of commuters is achieved voluntarily either by the commuters themselves, in order to save money, or by employers in order to meet some target stipulated by the planning authority. Use of car pooling can be promoted by allowing private cars with multiple occupants to use bus-lanes during peak hour travel or by allowing them reduced parking charges at their destination.

The car-centred approach

The car-centred approach has been favoured by a number of large cities within the US, most notably Los Angeles. It seeks to cater for future increases in traffic demand through the construction of bigger and better roads, be they inter-urban or intra-urban links. Such an approach usually involves prioritising the development of road linkages both within and between the major urban centres. Measures such as in-car information for drivers regarding points of congestion along their intended route and the installation of state-of-the-art traffic control technology at all junctions, help maximise usage along the available road space.

The public transport-centred approach

In the public transport-centred approach the strategy will emphasise the importance of bus and rail-based improvements as the preferred way of coping with increased transport demand. Supporters of this approach point to the environmental and social advantages of such a strategy, reducing noise and air pollution and increasing efficiency in the use of fossil fuels while also making transport available to those who cannot afford to run a car. However, the success of such a strategy depends on the ability of transport planners to induce increasing numbers of private car users to change their mode of travel during peak hours to public transport. This will minimise highway congestion as the number of peak hour journeys increase over the years. Such a result will only be achieved if the public transport service provided is clean, comfortable, regular and affordable.

1.4.4   Transportation studies

Whatever the nature of the proposed highway system under consideration, be it a new motorway to link two cities or a network of highway improvements within an urban centre, and whatever planning strategy the decision-makers are adopting (assuming that the strategy involves, to some extent, the construction of new/upgraded roadways), a study must be carried out to determine the necessity or appropriateness of the proposal. This process will tend to be divided into two subsections:

- A transportation survey to establish trip-making patterns
- The production and use of mathematical models both to predict future transport requirements and to evaluate alternative highway proposals.

Transportation survey

Initially, the responsible transport planners decide on the physical boundary within which the study will take place. Most transport surveys have at their basis the land-use activities within the study area and involve making an inventory of the existing pattern of trip making, together with consideration of the socio-economic factors that affect travel patterns. Travel patterns are determined by compiling a profile of the origin and destination (OD) of all journeys made within the study area, together with the mode of travel and the purpose of each journey. For those journeys originating within the study area, household surveys are used to obtain the OD information. These can be done with or without an interviewer assisting. In the case of the former, termed a personal interview survey, an interviewer records answers provided by the respondent. With the latter, termed a self-completion survey, the respondent completes a question-naire without the assistance of an interviewer, with the usual format involving the questionnaire being delivered/mailed out to the respondent who then mails it back/has it collected when all questions have been answered.

For those trips originating outside the study area, traversing its external 'cordon' and ending within the study area, the OD information is obtained by interviewing trip makers as they pass through the 'cordon' at the boundary of the study area. These are termed intercept surveys where people are intercepted in the course of their journey and asked where their trip started and where it will finish.

A transportation survey should also gather information on the adequacy of existing infrastructure, the land use activities within the study area and details on the socio-economic classification of its inhabitants. Traffic volumes along the existing road network together with journey speeds, the percentage of heavy goods vehicles using it and estimates of vehicle occupancy rates are usually required. For each designated zone within the study area, office and factory floor areas and employment figures will indicate existing levels of industrial/commercial activity, while census information and recommendations on housing densities will indicate population size. Some form of personal household-based survey will be required within each zone to determine household incomes and their effect on the frequency of trips made and the mode of travel used.

Production and use of mathematical models

At this point, having gathered all the necessary information, models are developed to translate the information on existing travel patterns and land-use profiles into a profile of future transport requirements for the study area. The four stages in constructing a transportation model are trip generation, trip distribution, modal split and traffic assignment. The first stage estimates the number of trips generated by each zone based on the nature and level of land-use activity within it. The second distributes these trips among all possible destinations, thus establishing a pattern of trip making between each of the zones. The mode of travel used by each trip maker to complete their journey is then determined and finally the actual route within the network taken by the trip maker in each case. Each of these four stages is described in detail in the next chapter. Together they form the process of transportation demand analysis which plays a central role within highway engineering. It attempts to describe and explain both existing and future travel behaviour in an attempt to predict demand for both car-based and other forms of transportation modes.

1.5   The decision-making process in highway and transport planning

1.5.1   Introduction

Highway and transportation planning can be described as a process of making decisions which concerns the future of a given transport system. The decisions relate to the determination of future demand; the relationships and interactions which exist between the different modes of transport; the effect of the proposed system on both existing land uses and those proposed for the future; the economic, environmental, social and political impacts of the proposed system and the institutional structures in place to implement the proposal put forward.

Transport planning is generally regarded as a rational process, i.e. a rational and orderly system for choosing between competing proposals at the planning stage of a project. It involves a combined process of information gathering and decision-making.

The five steps in the rational planning process are summarised in Table 1.3.

Table 1.3 Steps in the rational decision-making process for a transportation project

Step	Purpose
Definition of goals and objectives	To define and agree the overall purpose of the proposed transportation project
Formulation of criteria/measures of effectiveness	To establish standards of judging by which the transportation options can be assessed in relative and absolute terms
Generation of transportation alternatives	To generate as broad a range of feasible transportation options as possible
Evaluation of transportation alternatives	To evaluate the relative merit of each transportation option
Selection of preferred transportation alternative/group of alternatives	To make a final decision on the adoption of the most favourable transportation option as the chosen solution for implementation

In the main, transport professionals and administrators subscribe to the values underlying rational planning and utilise this process in the form detailed below. The rational process is, however, a subset of the wider political decision-making system, and interacts directly with it both at the goal-setting stage and at the point in the process at which the preferred option is selected. In both situations, inputs from politicians and political/community groupings representing those with a direct interest in the transport proposal under scrutiny are essential in order to maximise the level of acceptance of the proposal under scrutiny.

Assuming that the rational model forms a central part of transport planning and that all options and criteria have been identified, the most important stage within this process is the evaluation/appraisal process used to select the most appropriate transport option. Broadly speaking, there are two categories of appraisal process. The first consists of a group of methods that require the assessments to be solely in money terms. They assess purely the economic consequences of the proposal under scrutiny. The second category consists of a set of more widely-based techniques that allow consideration of a wide range of decision criteria – environmental, social and political as well as economic, with assessments allowable in many forms, both monetary and non-monetary. The former group of methods are termed economic evaluations, with the latter termed multi-criteria evaluations.

Evaluation of transport proposals requires various procedures to be followed. These are ultimately intended to clarify the decision relating to their approval. It is a vital part of the planning process, be it the choice between different location options for a proposed highway or the prioritising of different transport alternatives listed within a state, regional or federal strategy. As part of the process by which a government approves a highway scheme, in addition to the carrying out of traffic studies to evaluate the future traffic flows that the proposed highway will have to cater for, two further assessments are of particular importance to the overall approval process for a given project proposal:

- A monetary-based economic evaluation, generally termed a cost-benefit analysis (CBA)
- A multi-criteria-based environmental evaluation, generally termed an environmental impact assessment (EIA)

Layered on top of the evaluation process is the need for public participation within the decision process. Although a potentially time consuming procedure, it has the advantages of giving the planners an understanding of the public's concerns regarding the proposal and also actively draws all relevant interest groups into the decision-making system. The process, if properly conducted, should serve to give the decision-makers some reassurance that all those affected by the development have been properly consulted before the construction phase proceeds.

1.5.2   Economic assessment

Within the US, both economic and environmental evaluations form a central part of the regional transportation planning process called for by federal law when state level transportation plans required under the Intermodal Transportation Efficiency Act 1991 are being determined or in decisions by US federal organisations regarding the funding of discretionary programmes.

Cost-benefit analysis is the most widely used method of project appraisal throughout the world. Its origins can be traced back to a classic paper on the utility of public works by Dupuit (1844), written originally in the French language. The technique was first introduced in the US in the early part of the twentieth century with the advent of the Rivers and Harbours Act 1902 which required that any evaluation of a given development option must take explicit account of navigation benefits arising from the proposal, and these should be set against project costs, with the project only receiving financial support from the federal government in situations where benefits exceeded costs. Following this, a general primer, known as the 'Green Book', was prepared by the US Federal Interagency River Basin Committee (1950), detailing the general principles of economic analysis as they were to be applied to the formulation and evaluation of federally funded water resource projects. This formed the basis for the application of cost-benefit analysis to water resource proposals, where options were assessed on the basis of one criterion – their economic efficiency. In 1965 Dorfman released an extensive report applying cost-benefit analysis to developments outside the water resources sector. From the 1960s onwards the technique spread beyond the US and was utilised extensively to aid option choice in areas such as transportation.

Cost-benefit analysis is also widely used throughout Europe. The 1960s and 1970s witnessed a rapid expansion in the use of cost-benefit analysis within the UK as a tool for assessing major transportation projects. These studies included the cost-benefit analysis for the London Birmingham Motorway by Coburn Beesley and Reynolds (1960) and the economic analysis for the siting of the proposed third London airport by Flowerdew (1972). This growth was partly the result of the increased government involvement in the economy during the post-war period, and partly the result of the increased size and complexity of investment decisions in a modern industrial state. The computer programme COBA has been used since the early 1980s for the economic assessment of major highway schemes (DoT, 1982). It assesses the net value of a preferred scheme and can be used for determining the priority to be assigned to a specific scheme, for generating a shortlist of alignment options to be presented to local action groups for consultation purposes, or for the basic economic justification of a given corridor. In Ireland, the Department of Finance requires that all highway proposals are shown to have the capability of yielding a minimum economic return on investment before approval for the scheme will be granted.

Detailed information on the economic assessment of highway schemes is given in Chapter 3.

1.5.3   Environmental assessment

Any economic evaluation for a highway project must be viewed alongside its environmental and social consequences. This area of evaluation takes place within the environmental impact assessment (EIA) for the proposal. Within the US, EIA was brought into federal law under the National Environmental Policy Act 1969 which required an environmental assessment to be carried out in the case of all federally funded projects likely to have a major adverse effect on the quality of the human environment. This law has since been imposed at state level also.

Interest in EIA spread from America to Europe in the 1970s in response to the perceived deficiencies of the then existing procedures for appraising the environmental consequences of major development projects. The central importance of EIA to the proper environmental management and the prevention of pollution led to the introduction of the European Union Directive 85/337/EEC (Council of the European Communities, 1985) which required each member state to carry out an environmental assessment for certain categories of projects, including major highway schemes. Its overall purpose was to ensure that a mechanism was in place for ensuring that the environmental dimension is properly considered within a formal framework alongside the economic and technical aspects of the proposal at its planning stage.

Within the UK, the environmental assessment for a highway proposal requires 12 basic impacts to be assessed, including air, water and noise quality, landscape, ecology and land use effects, and impacts on culture and local communities, together with the disruption the scheme will cause during its construction. The relative importance of the impacts will vary from one project to another. The details of how the different types of impacts are measured and the format within which they are presented are given in Chapter 3.

1.5.4   Public consultation

For major trunk road schemes, public hearings are held in order to give interested parties an opportunity to take part in the process of determining both the basic need for the highway and its optimum location.

For federally funded highways in the US, at least one public hearing will be required if the proposal is seen to:

- Have significant environmental, social and economic effects
- Require substantial wayleaves/rights-of-way, *or*
- Have a significantly adverse effect on property adjoining the proposed highway.

Within the hearing format, the state highway agency representative puts forward the need for the proposed roadway, and outlines its environmental, social and economic impacts together with the measures put forward by them to mitigate, as far as possible, these effects. The agency is also required to take submissions from the public and consult with them at various stages throughout the project planning process.

Within the UK, the planning process also requires public consultation. Once the need for the scheme has been established, the consultation process centres on selecting the preferred route from the alternatives under scrutiny. In situations where only one feasible route can be identified, public consultation will still be undertaken in order to assess the proposal relative to the 'do-minimum' option. As part of the public participation process, a consultation document explaining the scheme in layman's terms and giving a broad outline of its cost and environmental/social consequences, is distributed to all those with a legitimate interest in the proposal. A prepaid questionnaire is usually included within the consultation document, which addresses the public's preferences regarding the relative merit of the alternative alignments under examination. In addition, an exhibition is held at all local council offices and public libraries at which the proposal is on public display for the information of those living in the vicinity of the proposal. Transport planners are obliged to take account of the public consultation process when finalising the chosen route for the proposed motorway. At this stage, if objections to this route still persist, a public enquiry is usually required before final approval is obtained from the secretary of state.

In Ireland, two public consultations are built into the project management guidelines for a major highway project. The first takes place before any alternatives are identified and seeks to involve the public at a preliminary stage in the scheme, seeking their involvement and general understanding. The second public consultation involves presentation of the route selection study and the recommended route, together with its likely impacts. The views and reactions of the public are recorded and any queries responded to. The route selection report is then reviewed in order to reflect any legitimate concerns of the public. Here also, the responsible government minister may determine that a public inquiry is necessary before deciding whether or not to grant approval for the proposed scheme.

1.6 Summary

Highway engineering involves the application of scientific principles to the planning, design, maintenance and operation of a highway project or system of projects. The aim of this book is to give students an understanding of the analysis and design techniques that are fundamental to the topic. To aid this, numerical examples are given throughout the book. This chapter has briefly introduced the context within which highway projects are undertaken, and details the frameworks, both institutional and procedural, within which the planning, design, construction and management of highway systems take place. The remainder of the chapters deal specifically with the basic technical details relating to the planning, design, construction and maintenance of schemes within a highway network.

Chapter 2 deals in detail with the classic four-stage model used to determine the volume of flow on each link of a new or upgraded highway network. The process of scheme appraisal is dealt with in Chapter 3, outlining in detail methodologies for both economic and environmental assessment and illustrating the format within which both these evaluations can be analysed. Chapter 4 demonstrates how the twin factors of predicted traffic volume and level of service to be provided by the proposed roadway determine the physical size and number of lanes provided. Chapter 5 details the basic design procedures for the three different types of highway intersections – priority junctions, roundabouts and signalised intersections. The fundamental principles of geometric design, including the determination of both vertical and horizontal alignments, are given in Chapter 6. Chapter 7 summarises the basic materials which comprise road pavements, both flexible and rigid, and outlines their structural properties, with Chapter 8 addressing details of their design and Chapter 9 dealing with their maintenance.

1.7   References

Coburn, T.M., Beesley, M.E. & Reynolds, D.J. (1960) *The London-Birmingham Motorway: Traffic and Economics*. Technical Paper No. 46. Road Research Laboratory, Crowthorne.

Council of the European Communities (1985) On the assessment of the effects of certain public and private projects on the environment. *Official Journal L175*, 28.5.85, 40–48 (85/337/EEC).

DoT (1982) Department of Transport *COBA: A method of economic appraisal of highway schemes*. The Stationery Office, London.

Dupuit, J. (1844) On the measurement of utility of public works. *International Economic Papers*, Volume 2.

Flowerdew, A.D.J. (1972) Choosing a site for the third London airport: The Roskill Commission approach. In R. Layard (ed.) *Cost-Benefit Analysis*. Penguin, London.

US Federal Interagency River Basin Committee (1950) Subcommittee on Benefits and Costs. *Proposed Practices for Economic Analysis of River Basin Projects*. Washington DC, USA.

Chapter 2

Forecasting Future Traffic Flows

2.1 Basic principles of traffic demand analysis

If transport planners wish to modify a highway network either by constructing a new roadway or by instituting a programme of traffic management improvements, any justification for their proposal will require them to be able to formulate some forecast of future traffic volumes along the critical links. Particularly in the case of the construction of a new roadway, knowledge of the traffic volumes along a given link enables the equivalent number of standard axle loadings over its lifespan to be estimated, leading directly to the design of an allowable pavement thickness, and provides the basis for an appropriate geometric design for the road, leading to the selection of a sufficient number of standard width lanes in each direction to provide the desired level of service to the driver. Highway demand analysis thus endeavours to explain travel behaviour within the area under scrutiny, and, on the basis of this understanding, to predict the demand for the highway project or system of highway services proposed.

The prediction of highway demand requires a unit of measurement for travel behaviour to be defined. This unit is termed a trip and involves movement from a single origin to a single destination. The parameters utilised to detail the nature and extent of a given trip are as follows:

- Purpose
- Time of departure and arrival
- Mode employed
- Distance of origin from destination
- Route travelled.

Within highway demand analysis, the justification for a trip is founded in economics and is based on what is termed the utility derived from a trip. An individual will only make a trip if it makes economic sense to do so, i.e. the economic benefit or utility of making a trip is greater than the benefit accrued by not travelling, otherwise it makes sense to stay at home as travelling results in no economic benefit to the individual concerned. Utility defines the 'usefulness' in economic terms of a given activity. Where two possible trips are open to an individual, the one with the greatest utility will be undertaken. The utility of any trip usually results from the activity that takes place at its destination. For example, for workers travelling from the suburbs into the city centre by car, the basic utility of that trip is the economic activity that it makes possible, i.e. the job done by the traveller for which he or she gets paid. One must therefore assume that the payment received by a given worker exceeds the cost of making the trip (termed disutility), otherwise it would have no utility or economic basis. The 'cost' need not necessarily be in money terms, but can also be the time taken or lost by the traveller while making the journey. If an individual can travel to their place of work in more than one way, say for example by either car or bus, they will use the mode of travel that costs the least amount, as this will allow them to maximise the net utility derived from the trip to their destination. (Net utility is obtained by subtracting the cost of the trip from the utility generated by the economic activity performed at the traveller's destination.)

2.2   Demand modelling

Demand modelling requires that all parameters determining the level of activity within a highway network must first be identified and then quantified in order that the results output from the model has an acceptable level of accuracy. One of the complicating factors in the modelling process is that, for a given trip emanating from a particular location, once a purpose has been established for making it, there are an enormous number of decisions relating to that trip, all of which must be considered and acted on simultaneously within the model. These can be classified as:

- Temporal decisions – once the decision has been made to make the journey, it still remains to be decided when to travel
- Decisions on chosen journey destination – a specific destination must be selected for the trip, e.g. a place of work, a shopping district or a school
- Modal decisions – relate to what mode of transport the traveller intends to use, be it car, bus, train or slower modes such as cycling/walking
- Spatial decisions – focus on the actual physical route taken from origin to final destination. The choice between different potential routes is made on the basis of which has the shorter travel time.

If the modelling process is to avoid becoming too cumbersome, simplifications to the complex decision-making processes within it must be imposed. Within a basic highway model, the process of simplification can take the form of two stages:

(1) Stratification of trips by purpose and time of day
(2) Use of separate models in series for estimating the number of trips made from a given geographical area under examination, the origin and destination of each, the mode of travel used and the route selected.

Stratification entails modelling the network in question for a specific time of the day, most often the morning peak hour but also, possibly, some critical off-peak period, with trip purpose being stratified into work and non-work. For example, the modeller may structure the choice sequence where, in the first instance, all work-related trips are modelled during the morning peak hour. (Alternatively, it may be more appropriate to model all non-work trips at some designated time period during the middle of the day.) Four distinct traffic models are then used sequentially, using the data obtained from the stratified grouping under scrutiny, in order to predict the movement of specific segments of the area's population at a specific time of day. The models are described briefly as:

- The trip generation model, estimating the number of trips made to and from a given segment of the study area
- The trip distribution model, estimating the origin and destination of each trip
- The modal choice model, estimating the form of travel chosen for each trip
- The route assignment model, predicting the route selected for each trip.

Used in series, these four constitute what can be described as the basic travel demand model. This sequential structure of traveller decisions constitutes a considerable simplification of the actual decision process where all decisions related to the trip in question are considered simultaneously, and it provides a sequence of mathematical models of travel behaviour capable of meaningfully forecasting traffic demand.

An overall model of this type may also require information relating to the prediction of future land uses within the study area, along with projections of the socio-economic profile of the inhabitants, to be input at the start of the modelling process. This evaluation may take place within a land use study.

Figure 2.1 illustrates the sequence of a typical transport demand model.

At the outset, the study area is divided into a number of geographical segments or zones. The average set of travel characteristics for each zone is then determined, base on factors such as the population of the zone in question. This grouping removes the need to measure each inhabitant's utility for travel, a task which would in any case from the modeller's perspective be virtually impossible to achieve.

The ability of the model to predict future travel demand is based on the assumption that future travel patterns will resemble those of the past. Thus the model is initially constructed in order to predict, to some reasonable degree of accuracy, present travel behaviour within the study area under scrutiny. Information on present travel behaviour within the area is analysed to determine meaningful regression coefficients for the independent variables that will predict the dependent variable under examination. This process of calibration will generate an equation where, for example, the existing population of a zone, multiplied by the appropriate coefficient, added to the average number of workers at present per household multiplied by its coefficient, will provide the number of

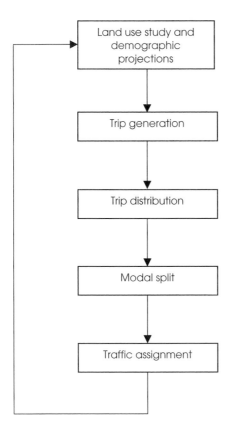

Figure 2.1 Sequence of transport demand model.

work trips currently originating from the zone in question. Once the modeller is satisfied that the set of values generated by the process is realistic, the calibration stage can be completed and the prediction of trips originating from the zone in question at some point in the future can be estimated by changing the values of the independent variables based on future estimates from experts.

2.3 Land use models

The demand for movement or trip making is directly connected to the activities undertaken by people. These activities are reflected in both the distribution and type of land uses within a given area. By utilising relationships between present day land uses and consequent movements in a given area, estimates of future movements given on land-use projections can be derived. The derivation of relationships between land uses and people movements are thus fundamental to an effective transport planning process. A land use model will thus estimate the future development for each of the zones within the study area, with estimates relating not only to predictions regarding the different land uses but also to those socio-economic variables that form the basic data for trip generation, the first of the four-stage sequential models. Input by experienced land-use planners is essential to the success of this phase. The end product of the land-use forecasting process usually takes the form of a land use plan where land-use estimates stretching towards some agreed time horizon, usually between 5 and 25 years, are agreed.

The actual numerical relationship between land use and movement information is derived using statistical/mathematical techniques. A regression analysis is employed to establish, for a given zone within the study area, the relationship between the vehicle trips produced by or attracted to it and characteristics derived both from the land use study and demographic projections. This leads us on directly to the first trip modelling stage – trip generation.

2.4   Trip generation

Trip generation models provide a measure of the rate at which trips both in and out of the zone in question are made. They predict the total number of trips produced by and attracted to its zone. Centres of residential development, where people live, generally produce trips. The more dense the development and the greater the average household income is within a given zone, the more trips will be produced by it. Centres of economic activity, where people work, are the end point of these trips. The more office, factory and shopping space existing within the zone, the more journeys will terminate within it. These trips are 2-way excursions, with the return journey made at some later stage during the day.

It is an innately difficult and complex task to predict exactly when a trip will occur. This complexity arises from the different types of trips that can be undertaken by a car user during the course of the day (work, shopping, leisure, etc.). The process of stratification attempts to simplify the process of predicting the number and type of trips made by a given zone. Trips are often stratified by purpose, be it work, shopping or leisure. Different types of trips have different characteristics that result in them being more likely to occur at different times of the day. The peak time for the journey to work is generally in the early morning, while shopping trips are most likely during the early evening. Stratification by time, termed temporal aggregation, can also be used, where trip generation models predict the number of trips per unit timeframe during any given day. An alternative simplification procedure can involve considering the trip behaviour of an entire household of travellers rather than each individual trip maker within it. Such an approach is justified by the homogeneous nature, in social and economic terms, of the members of a household within a given zone.

Within the context of an urban transportation study, three major variables govern the rate at which trips are made from each zone within the study area:

- Distance of zone from the central business district/city centre area
- Socio-economic characteristics of the zone population (per capita income, cars available per household)
- Intensity of land use (housing units per hectare, employees per square metre of office space).

The relationships between trips generated and the relevant variables are expressed as mathematical equations, generally in a linear form. For example, the model could take the following form:

$$T_{ij} = \alpha_0 + \alpha_1 Z_{1j} + \alpha_2 Z_{2j} + \cdots + \alpha_n Z_{nj} \tag{2.1}$$

where

T_{ij} = number of vehicle trips per time period for trip type i (work, non-work) made by household j

Z = characteristic value n for household j, based on factors such as the household income level and number of cars available within it

α = regression coefficient estimated from travel survey data relating to n

A typical equation obtained for a transportation study in the UK might be:

$$T = 0 + 0.07 * Z_1 + 0.005 * Z_2 + 0.95 * Z_3 - 0.003 * Z_4$$

where

T = total number of trips per household per 24 hours
Z_1 = family size
Z_2 = total income of household
Z_3 = cars per household
Z_4 = housing density

Example 2.1 – Basic calculation of trip rates

The following model is compiled for shopping trips generated during the weekly peak hour for this activity (5.30 PM to 6.30 PM on Fridays). The relationship is expressed as follows:

$$T_{shopping} = 0.15 + 0.1 * Z_1 + 0.01 * Z_2 - 0.145 * Z_3$$

where

T = total number of vehicle-based shopping trips per household in peak hour
Z_1 = household size
Z_2 = annual income of household (in £000s)
Z_3 = employment in neighbourhood (in 00s)

Calculate the trip rate for a household of four people with an annual income of £30 000 within a neighbourhood where 1000 people are employed.

Solution

$$\text{Number of trips} = 0.15 + 0.1 * 4 + 0.01 * 30 - 0.145 * 10$$
$$= 2.3 \text{ vehicle trips}$$

(The negative sign in the above equation arises from the reduced likelihood of a non-work related trip occurring within an area of high employment.)

The coefficients α_0 to α_n which occur within typical trip generation models as shown in equation 2.1 are determined through regression analysis. Manual solutions from multiple regression coefficients can be tedious and time-consuming but software packages are readily available for solving them. For a given trip generation equation, the coefficients can be assumed to remain constant over time for a given specified geographical location with uniform demographic and socio-economic factors.

In developing such regression equations, among the main assumptions made is that all the variables on the right-hand side of the equation are independent of each other. It may not, however, be possible for the transportation expert to conform to such a requirement and this may leave the procedure open to a certain level of criticism. In addition, basic errors in the regression equation may exist as a result of biases or inaccuracies in the survey data from which it was derived. Equation 2.1 assumes that the regression of the dependent variable on the independent variables is linear, whereas in reality this may not be the case.

Difficulties with the use of regression analysis for the analysis of trip generations have resulted in support for the use of models with the person or, more often, the household, at its basis. This process of estimating trip generations directly from household data is known as category analysis. Within it, households are subdivided into smaller groupings that are known to possess set trip-making patterns. Category analysis assumes that the volume of trips generated depend on the characteristics of households together with their location relative to places of work. These characteristics are easily measured. They include household income, car ownership, family size, number of workers within the household and housing density. The method does, however, assume that both car ownership and real income levels will increase in the future. This may not necessarily be the case.

For example, the more people within a household and the more cars available to them, the more trips they will make; say we define 15 subgroups in terms of two characteristics – numbers within the household and number of cars available – and we estimate the number of trips each subgroup is likely to make during the course of the day. An example of category analysis figures is given in Table 2.1.

Household pop.	Available cars per household		
	0	1	2+
1	1.04	1.85	2.15
2	2.02	3.10	3.80
3	2.60	3.40	4.00
4	3.80	4.80	6.40
5+	4.20	5.20	6.40

Table 2.1 Category analysis table (daily trip rates per household category)

For the neighbourhood under examination, once the number of households within each subgroup is established, the total number trips generated each day can be calculated.

Example 2.2 – Calculating trip rates using category analysis
For a given urban zone, using the information on trip rates given in Table 2.1 and the number of each household category within it as given in Table 2.2, calculate the total number of daily trips generated by the 100 households within the zone.

Solution

For each table cell, multiply the trip rate for each category by the number of households in each category, summing all values to obtain a total number of daily trips as follows:

$$T = 4*1.04 + 23*1.85 + 2*2.15 + 2*2.02 + 14*3.1 + 14*3.8 + 1*2.6$$
$$+ 9*3.4 + 14*4.0 + 0*3.8 + 5*4.8 + 7*6.4 + 0*4.2 + 1*5.2 + 4*6.4$$
$$= 340.45$$

Household pop.	Available cars per household		
	0	1	2+
1	4	23	2
2	2	14	14
3	1	9	14
4	0	5	7
5+	0	1	4

Table 2.2 Category analysis table (number of households from within zone in each category, total households = 100)

2.5   Trip distribution

2.5.1   *Introduction*

The previous model determined the number of trips produced by and attracted to each zone within the study area under scrutiny. For the trips produced by the zone in question, the trip distribution model determines the individual zones where each of these will end. For the trips ending within the zone under examination, the individual zone within which each trip originated is determined. The model thus predicts zone-to-zone trip interchanges. The process connects two known sets of trip ends but does not specify the precise route of the trip or the mode of travel used. These are determined in the two last phases of the modelling process. The end product of this phase is the formation of a trip matrix

Zone of origin	Zone of destination				
	1	2	3	4	....
1	T_{11}	T_{12}	T_{13}	T_{14}	....
2	T_{21}	T_{22}	T_{23}	T_{24}	....
3	T_{31}	T_{32}	T_{33}	T_{34}	....
4	T_{41}	T_{42}	T_{43}	T_{44}	....
.	.	.	.	.	
.	.	.	.	.	
.	.	.	.	.	

Table 2.3 Origin-destination matrix (e.g. T_{14} = number of trips originating in zone 1 and ending in zone 4)

between origins and destinations, termed an origin-destination matrix. Its layout is illustrated in Table 2.3.

There are several types of trip distribution models, including the gravity model and the Furness method.

2.5.2   *The gravity model*

The gravity model is the most popular of all the trip distribution models. It allows the effect of differing physical planning strategies, travel costs and transportation systems to be taken into account. Within it, existing data is analysed in order to obtain a relationship between trip volumes and the generation and attraction of trips along with impedance factors such as the cost of travel.

The name is derived from its similarity to the law of gravitation put forward by Newton where trip interchange between zones is directly proportional to the attractiveness of the zones to trips, and inversely proportional to some function of the spatial separation of the zones.

The gravity model exists in two forms:

$$T_{ij} = \frac{P_i A_j F_{ij}}{\sum_j (A_j F_{ij})} \tag{2.2}$$

or

$$T_{ij} = \frac{A_j P_i F_{ij}}{\sum_j (P_i F_{ij})} \tag{2.3}$$

where
T_{ij} = trips from zone i to zone j
A_j = trip attractions in zone j
P_i = trip productions in zone i
F_{ij} = impedance of travel from zone i to zone j

The impedance term, also called the deterrence function, refers to the resistance associated with the travel between zone *i* and zone *j* and is generally taken as a function of the cost of travel, travel time or travel distance between the two zones in question. One form of the deterrence function is:

$$F_{ij} = C_{ij}^{-\alpha} \tag{2.4}$$

The impedance function is thus expressed in terms of a generalised cost function C_{ij} and the α term which is a model parameter established either by analysing the frequency of trips of different journey lengths or, less often, by calibration.

Calibration is an iterative process within which initial values for Equation 2.4 are assumed and Equation 2.2 or 2.3 is then calculated for known productions, attractions and impedances computed for the baseline year. The parameters within Equation 2.4 are then adjusted until a sufficient level of convergence is achieved.

Example 2.3 – Calculating trip distributions using the gravity model
Taking the information from an urban transportation study, calculate the number of trips from the central business zone (zone 1) to five other surrounding zones (zone 2 to zone 6).

Table 2.4 details the trips produced by and attracted to each of the six zones, together with the journey times between zone 1 and the other five zones.

Use Equation 2.2 to calculate the trip numbers. Within the impedance function, the generalised cost function is expressed in terms of the time taken to travel between zone 1 and each of the other five zones and the model parameter is set at 1.9.

Solution

Taking first the data for journeys between zone 1 and zone 2, the number of journeys attracted to zone 2, A_2, is 45 000. The generalised cost function for the journey between the two zones is expressed in terms of the travel time between them: 5 minutes. Using the model parameter value of 1.9, the deterrence function can be calculated as follows:

$$F_{12} = 1 \div 5^{(1.9)} = 0.047$$

This value is then multiplied by A_2:

$$A_2 \times F_{12} = 2114$$

Summing $(A_j \times F_{1j})$ for $j = 2 \rightarrow 6$ gives a value of 2114 (see Table 2.5)

This value is divided into A_2, and multiplied by the number of trips produced by zone 1 (P_1) to yield the number of trips predicted to take place from zone 1 to zone 2, i.e.

Contd

Example 2.3 *Contd*

$$T_{12} = P_1 \times \left[(A_2 \times F_{12}) \div \sum (A_j \times F_{1j}) \right]$$
$$= 10000 \times (2114 \div 2597)$$
$$= 8143$$

Table 2.5 details the sequence involved in the calculation of all five trip volumes, $T_{12}, T_{13}, T_{14}, T_{15}, T_{16}$.

Zone	Generalised cost (travel time in mins.)	Productions	Attractions
1		10 000	15 000
2	5	7 500	45 000
3	10	15 000	25 000
4	15	12 500	12 500
5	20	8 000	15 000
6	25	5 000	20 000

Table 2.4 Trip productions, attractions and travel times between zones

Zone	A_j	$C_{i,j}$	$F_{i,j}$	$A_j F_{i,j}$	$\dfrac{A_j F_{ij}}{\sum_j (F_{ij})}$	$T_{1,j}$
1	15 000					
2	45 000	5	0.047	2114	0.814	8143
3	25 000	10	0.013	315	0.121	1212
4	12 500	15	0.006	73	0.028	280
5	15 000	20	0.003	51	0.020	195
6	20 000	25	0.002	44	0.017	170
Σ				2597	1.000	10 000

Table 2.5 Estimation of trip volumes between zone 1 and zones 2 to 6

As illustrated by Equations 2.2 and 2.3, the gravity model can be used to distribute either the productions from zone i or the attractions to zone j. If the calculation shown in Example 2.1 is carried out for the other five zones so that T_{2j}, T_{3j}, T_{4j}, T_{5j} and T_{6j} are calculated, a trip matrix will be generated with the rows of the resulting interchange matrix always summing to the number of trips produced within each zone because of the form of Equation 2.2. However, the columns when summed will not give the correct number of trips attracted to each zone. If, on the other hand, Equation 2.3 is used, the columns will sum correctly whereas the rows will not. In order to generate a matrix where row and column values sum correctly, regardless of which model is used, an iterative correction procedure, termed the row–column factor technique, can be used. This technique is demonstrated in the final worked example in section 2. It is explained briefly here.

Assuming Equation 2.2 is used, the rows will sum correctly but the columns will not. The first iteration of the corrective procedure involves each value of T_{ij} being modified so that each column will sum to the correct total of attractions.

$$T'_{ij} = \frac{A_j}{\sum_j T_{ij}}$$
(2.5)

Following this initial procedure, the rows will no longer sum correctly. Therefore, the next iteration involves a modification to each row so that they sum to the correct total of trip productions.

$$T'_{ij} = \frac{P_i}{\sum_i T_{ij}}$$
(2.6)

This sequence of corrections is repeated until successive iterations result in changes to values within the trip interchange matrix less than a specified percentage, signifying that sufficient convergence has been obtained. If Equation 2.3 is used, a similar corrective procedure is undertaken, but in this case the initial iteration involves correcting the production summations.

2.5.3   Growth factor models

The cells within a trip matrix indicate the number of trips between each origin-destination pair. The row totals give the number of origins and the column totals give the number of destinations. Assuming that the basic pattern of traffic does not change, traffic planners may seek to update the old matrix rather than compile a new one from scratch. The most straightforward way of doing this is by the application of a uniform growth factor where all cells within the existing matrix are multiplied by the same value in order to generate an updated set of figures.

$$T'_{ij} = T^t_{ij} \times G^{tt'}$$
(2.7)

where
T'_{ij} = Trips from zone i to zone j in some future forecasted year t'
T^t_{ij} = Trips from zone i to zone j in the present year under observation t
$G^{tt'}$ = Expected growth in trip volumes between year t and year t'

One drawback of this approach lies in the assumption that all zones will grow at the same rate. In reality, it is likely that some will grow at a faster rate than others. An approach that allows for such situations is the singly-constrained growth factor approach, which can be applied to either origin or destination data, but not both. The former application is termed the origin constrained growth factor method where a specific growth factor is applied to all trips originating in zone i (see Equation 2.8 below), while the latter is termed the destination constrained growth factor method where a specific growth factor is applied to all trips terminating in zone j (see Equation 2.9 below).

$$T_{ij}^{t'} = T_{ij}^{t} \times G_i^{tt'} \tag{2.8}$$

$$T_{ij}^{t'} = T_{ij}^{t} \times G_j^{tt'} \tag{2.9}$$

where

$G_i^{tt'}$ = Expected growth in trip volumes between year t and year t' for trips with their origin in zone i (*origins only*)

$G_j^{tt'}$ = Expected growth in trip volumes between year t and year t' for trips with their destination in zone j (*destinations only*)

Where information exists on zone-specific growth factors for both origins and destinations an average factor method can be applied where, for each origin-destination pair, the overall zone-specific growth factor is obtained from the average of the expected growth from origin i and destination j:

$$T_{ij}^{t'} = T_{ij}^{t} \times \left[\frac{G_i^{tt'} + G_j^{tt'}}{2} \right] \tag{2.10}$$

To obtain a more precise answer, however, a doubly constrained growth factor method can be used. One of the most frequently used models of this type was devised by K.P. Furness (the Furness method).

2.5.4   *The Furness method* (Furness, 1965)

This again is a growth factor method, but in this instance the basic assumption is that in the future the pattern of trip making will remain substantially identical to those at present, with the trip volumes increasing in line with the growth of both the generating *and* attracting zones. It is still more straightforward than the gravity model and quite applicable to situations where substantial changes in external factors such as land use are not expected.

The basic information required in order to initiate this procedure can be summarised as:

Data

T_{ij}^{t} – The existing trip interchange matrix (in baseline year t)

O_i – The total number of trips predicted to start from zone i in the future forecasted year

D_j – The total number of trips predicted to terminate in zone j in the future forecasted year.

To be computed

$T_{ij}^{t'}$ – The revised trip interchange matrix (in forecasted year t')

$G_i^{tt'}$ – Origin growth factor for row i (growth between year t and year t')

$G_j^{tt'}$ – Destination growth factor for column j (growth between year t and year t').

The sequence involved in the Furness method is:

(1)   The origin growth factor is calculated for each row of the trip interchange matrix using the following formula

$$G_i^{tt'} = O_i \Big/ \sum\nolimits_j T_{ij}^t \qquad\qquad (2.11)$$

(2)   Check whether the origin growth factors are within approximately 5% of unity. If they are, the procedure is not required. If they are not, proceed to the next step

(3)   Multiply the cells in each column of T_{ij}^t by its origin growth factor $G_i^{tt'}$ to produce the first version of the revised matrix $T_{ij}^{t'}$

(4)   The destination growth factor is calculated for each column of the trip interchange matrix using the following formula:

$$G_j^{tt'} = D_j \Big/ \sum\nolimits_i T_{ij}^{t'} \qquad\qquad (2.12)$$

(5)   Check whether the destination growth factors are within approximately 5% of unity. If they are, the procedure is not required. If they are not, proceed to the next step

(6)   Multiply the cells in each row of the first version of $T_{ij}^{t'}$ by its destination growth factor $G_i^{tt'}$ to produce the second version of $T_{ij}^{t'}$

(7)   Recalculate the origin growth factor:

$$G_i^{tt'} = O_i \Big/ \sum\nolimits_j T_{ij}^{t'} \qquad\qquad (2.13)$$

(8)   Proceed back to point 2.

(9)   Repeat the process until both the origin or destination growth factors being calculated are sufficiently close to unity (within 5% is usually permissible).

Example 2.4 – Furness method of trip distribution
Table 2.6 gives the matrix of present flows to and from four zones within a transportation study area. It also provides the total number of trips predicted to start from zone i, and the total number of trips predicted to terminate in zone j. Calculate the final set of distributed flows to and from the four zones.

Solution

Table 2.7 gives the origin and destination growth factors.
    Table 2.8 multiplies all the trip cells by the appropriate origin growth factors and a new set of destination growth factors are estimated. These are well outside unity.
    Table 2.9 multiplies all trip volumes in Table 2.8 by the amended destination growth factors to give a new matrix. From these a new set of origin growth factors are estimated. The factors are still not within 5% of unity.
    Tables 2.10 to 2.13 repeat the above sequence until the factors are seen to be within 5% of unity.

Contd

Example 2.4 Contd

Table 2.6 Matrix of existing flows and forecasted outbound and inbound trip totals

Origin	Destination				Forecasted total origins
	1	2	3	4	
1	0	300	750	225	3825
2	150	0	450	75	1675
3	300	300	0	450	2100
4	150	120	600	0	1375
Forecasted total destinations	700	1000	5500	1800	

Table 2.7 Calculation of origin and destination growth factors

Origin	Destination				Existing total origins	Forecasted total origins	Origin growth factor
	Z1	Z2	Z3	Z4			
Z1	0	300	750	225	1275	3825	3.00
Z2	150	0	450	75	675	1675	2.48
Z3	300	300	0	450	1050	2100	2.00
Z4	150	120	600	0	870	1375	1.58
Existing total destinations	600	720	1800	750			
Forecasted total destinations	700	1000	5500	1800			
Destination growth factor	1.17	1.39	3.06	2.4			

Table 2.8 Production of first amended matrix and revision of destination growth factors

Origin	Destination			
	Z1	Z2	Z3	Z4
Z1	0	900	2250	675
Z2	372	0	1117	186
Z3	600	600	0	900
Z4	237	190	948	0
Amended destination flows	1209	1690	4315	1761
Forecasted destination flows	700	1000	5500	1800
Destination growth factor	0.58	0.59	1.27	1.02

Table 2.9 Production of second revised matrix and revision of origin growth factors

Origin	Destination				Amended outbound flows	Forecasted outbound flows	Growth factor
	Z1	Z2	Z3	Z4			
Z1	0	533	2868	690	4091	3825	0.94
Z2	215	0	1423	190	1828	1675	0.92
Z3	347	355	0	920	1622	2100	1.29
Z4	137	112	1209	0	1458	1375	0.94

Contd

Example 2.4 Contd

Origin	Destination				
	Z1	Z2	Z3	Z4	
Z1	0	498	2682	645	
Z2	197	0	1303	174	
Z3	450	460	0	1191	
Z4	129	106	1140	0	
Amended inbound flows	776	1064	5125	2010	
Forecasted inbound flows	700	1000	5500	1800	
Destination growth factor	0.90	0.94	1.07	0.90	

Table 2.10 Production of third revised matrix and further revision of destination growth factors

Table 2.11 Production of fourth revised matrix and further revision of origin growth factors

Origin	Destination				Amended outbound flows	Forecasted outbound flows	Growth factor
	Z1	Z2	Z3	Z4			
Z1	0	468	2878	578	3924	3825	0.98
Z2	178	0	1399	156	1733	1675	0.97
Z3	405	432	0	1066	1903	2100	1.10
Z4	117	100	1223	0	1440	1375	0.96

Origin	Destination			
	Z1	Z2	Z3	Z4
Z1	0	456	2805	563
Z2	172	0	1352	151
Z3	447	477	0	1176
Z4	111	95	1168	0
Amended inbound flows	730	1028	5325	1890
Forecasted inbound flows	700	1000	5500	1800
Destination growth factor	0.96	0.97	1.03	0.95

Table 2.12 Production of fifth revised matrix and further revision of destination growth factors

Table 2.13 Production of sixth revised matrix and final required revision of origin growth factors (sufficient convergence obtained)

Origin	Destination				Amended total origins	Forecasted total origins	Growth factor
	Z1	Z2	Z3	Z4			
Z1	0	444	2897	536	3877	3825	0.987
Z2	165	0	1396	144	1705	1675	0.983
Z3	428	464	0	1120	2012	2100	1.044
Z4	107	92	1207	0	1406	1375	0.978

The use of growth factor methods such as the Furness technique is, to a large extent, dependent on the precise estimation of the actual growth factors used. These are a potential source of significant inaccuracy. The overriding drawback of these techniques is the absence of any measure of travel impedance. They cannot therefore take into consideration the effect of new or upgraded travel facilities or the negative impact of congestion.

2.6   Modal split

Trips can be completed using different modes of travel. The proportion of trips undertaken by each of the different modes is termed modal split. The simplest form of modal split is between public transport and the private car. While modal split can be carried out at any stage in the transportation planning process, it is assumed here to occur between the trip distribution and assignment phases. The trip distribution phase permits the estimation of journey times/costs for both the public and private transport options. The modal split is then decided on the basis of these relative times/costs. In order to simplify the computation of modal split, journey time is taken as the quantitative measure of the cost criterion.

The decision by a commuter regarding choice of mode can be assumed to have its basis in the micro-economic concept of utility maximisation. This model presupposes that a trip maker selects one particular mode over all others on the basis that it provides the most utility in the economic sense. One must therefore be in a position to develop an expression for the utility provided by any one of a number of mode options. The function used to estimate the total utility provided by a mode option usually takes the following form:

$$U_m = \beta_m + \sum \alpha_j z_{mj} + \varepsilon \qquad (2.14)$$

where
U_m = total utility provided by mode option m
β_m = mode specific parameter
z_{mj} = set of travel characteristics of mode m, such as travel time or costs
α_j = parameters of the model, to be determined by calibration from travel survey data
ε = stochastic term which makes allowance for the unspecifiable portion of the utility of the mode that is assumed to be random

The β_m terms state the relative attractiveness of different travel modes to those within the market segment in question. They are understood to encapsulate the effect of all the characteristics of the mode not incorporated within the z terms. The 'ε' term expresses the variability in individual utilities around the average utility of those within the market segment.

Based on these definitions of utility, the probability that a trip maker will select one mode option, m, is equal to the probability that this option's utility is greater than the utility of all other options. The probability of a commuter choosing mode m (bus, car, train) can thus be represented by the following multi-nomial logit choice model:

$$P_m = \frac{e^{(um)}}{\sum e^{(um')}} \tag{2.15}$$

where
P_m = probability that mode m is chosen
m' = index over all modes included in chosen set

Details of the derivation of Equation 2.15 are provided in McFadden (1981).

Where only two modes are involved, the above formula simplifies to the following binary logit model:

$$P_1 = \frac{1}{1 + e^{(u_2 - u_1)}} \tag{2.16}$$

Example 2.5 – Use of multi-nomial logit model for estimation of modal split
Use a logit model to determine the probabilities of a group of 5000 work commuters choosing between three modes of travel during the morning peak hour:

- Private car
- Bus
- Light rail.

The utility functions for the three modes are estimated using the following equations:

$$U_C = 2.4 - 0.2C - 0.03T$$
$$U_B = 0.0 - 0.2C - 0.03T$$
$$U_{LR} = 0.4 - 0.2C - 0.03T$$

where
C = cost (£)
T = travel time (minutes)

For all workers:

- The cost of driving is £4.00 with a travel time of 20 minutes
- The bus fare is £0.50 with a travel time of 40 minutes
- The rail fare is £0.80 with a travel time of 25 minutes.

Contd

Example 2.5 Contd

Solution

Substitute costs and travel times into the above utility equations as follows:

U_C = 2.4 − 0.2 (4) − 0.03 (20)   = 1.00
U_B = 0.0 − 0.2 (0.5) − 0.03 (40) = −1.30
U_{LR} = 0.4 − 0.2 (0.8) − 0.03 (25) = −0.51

$e^{1.0}$ = 2.7183
$e^{-1.3}$ = 0.2725
$e^{-0.51}$ = 0.6005

P_{CAR} = 2.7183/(2.7183 + 0.2725 + 0.6005) = 0.757 (75.7%)
P_{BUS} = 0.2725/(2.7183 + 0.2725 + 0.6005) = 0.076 (7.6%)
P_{RAIL} = 0.6005/(2.7183 + 0.2725 + 0.6005) = 0.167 (16.7%)

Thus, 3785 commuters will travel to work by car, 380 by bus and 835 by light rail.

Example 2.6 − Effect of introducing bus lane on modal split figures
Taking a suburban route with the same peak hour travel conditions for car and bus as described in Example 2.5, the local transport authority constructs a bus lane in order to alter the modal split in favour of bus usage. When in operation, the bus lane will reduce the bus journey time to 20 minutes and will increase the car travel time to 30 minutes. The cost of travel on both modes remains unaltered.

Calculate the modal distributions for the 1000 work commuters using the route both before and after the construction of the proposed new bus facility.

Solution

The baseline utilities for the two modes are as in Example 2.5:

U_C = 1.00
U_B = −1.30

The modal distributions are thus:

Contd

Example 2.6 Contd

$$P_{CAR} = e^{1.0}/(e^{1.0} + e^{-1.3}) = 0.91 \ (91\%)$$
$$P_{BUS} = e^{-1.3}/(e^{1.0} + e^{-1.3}) = 0.09 \ (9\%)$$

These probabilities can also be calculated using Equation 2.16:

$$P_{CAR} = 1/(1 + (e^{(-1.3-1.0)})) = 0.91 \ (91\%)$$
$$P_{BUS} = 1/(1 + (e^{(1.0-(-1.3))})) = 0.09 \ (9\%)$$

During the morning peak hour, 910 commuters will therefore travel by car with the remaining 90 taking the bus.

After construction of the new bus lane, the changed journey times alter the utilities as follows:

$$U_C = 2.4 - 0.2(4) - 0.03(30) = 0.70$$
$$U_B = 0.0 - 0.2(0.5) - 0.03(20) = -0.70$$

Based on these revised figures, the new modal splits are:

$$P_{CAR} = e^{0.7}/(e^{0.7} + e^{-0.7}) = 0.80 \ (80\%)$$
$$P_{BUS} = e^{-0.7}/(e^{0.7} + e^{-0.7}) = 0.20 \ (20\%)$$

Post construction of the bus lane, during the morning peak hour, 800 (−110) commuters will now travel by car with 200 (+ 110) taking the bus.

Thus, the introduction of the bus lane has more than doubled the number of commuters travelling by bus.

2.7   Traffic assignment

Traffic assignment constitutes the final step in the sequential approach to traffic forecasting. The output from this step in the process will be the assignment of precise quantities of traffic flow to specific routes within each of the zones.

Assignment requires the construction of a mathematical relationship linking travel time to traffic flow along the route in question. The simplest approach involves the assumption of a linear relationship between travel time and speed on the assumption that free-flow conditions exist, i.e. the conditions a trip maker would experience if no other vehicles were present to hinder travel speed. In this situation, travel time can be assumed to be independent of the volume of traffic using the route. (The 'free-flow' speed used assumes that vehicles travel along the route at the designated speed limit.) A more complex parabolic speed/flow relationship involves travel time increasing more quickly as traffic flow reaches capacity. In this situation, travel time *is* volume dependent.

In order to develop a model for route choice, the following assumptions must be made:

(1) Trip makers choose a route connecting their origin and destination on the basis of which one gives the shortest travel time

(2) Trip makers know the travel times on all available routes between the origin and destination.

If these two assumptions are made, a rule of route choice can be assembled which states that trip makers will select a route that minimises their travel time between origin and destination. Termed Wardrop's first principle, the rule dictates that, on the assumption that the transport network under examination is at equilibrium, individuals cannot improve their times by unilaterally changing routes (Wardrop, 1952). If it is assumed that travel time is independent of the traffic volume along the link in question, all trips are assigned to the route of minimum time/cost as determined by the 'all-or-nothing' algorithm illustrated in Example 2.7.

Example 2.7 – The 'all-or nothing' method of traffic assignment
The minimum time/cost paths for a six-zone network are given in Table 2.14, with the average daily trip interchanges between each of the zones given in Table 2.15.

Using the 'all-or-nothing' algorithm, calculate the traffic flows on each link of the network.

Solution

For each of the seven links in the network, (1-2, 1-4, 2-3, 2-5, 3-6, 4-5, 5-6), the pairs contributing to its total flow are:

Link 1-2: flows from 1-2, 2-1, 1-3, 3-1, 1-5, 5-1, 1-6, 6-1
Link 1-4: flows from 1-4, 4-1
Link 2-3: flows from 1-3, 3-1, 1-6, 6-1, 3-4, 4-3, 2-3, 3-2, 3-5, 5-3
Link 2-5: flows from 1-5, 5-1, 2-4, 4-2, 4-3, 3-4, 2-5, 5-2, 2-6, 6-2, 3-5, 5-3
Link 3-6: flows from 1-6, 6-1, 3-6, 6-3
Link 4-5: flows from 4-2, 2-4, 4-5, 5-4, 4-3, 3-4, 4-6, 6-4
Link 5-6: flows from 4-6, 6-4, 2-6, 6-2, 5-6, 6-5

The link flows can thus be computed as:

Link flow 1-2: 250 + 300 + 150 + 100 + 100 + 150 + 75 + 150 = 1275
Link flow 1-4: 125 + 200 = 325
Link flow 2-3: 100 + 150 + 75 + 150 + 50 + 100 + 275 + 325 + 125 + 100 = 1450
Link flow 2-5: 150 + 100 + 150 + 200 + 50 + 100 + 400 + 300 + 150 + 150 + 125 + 100 = 1975
Link flow 3-6: 75 + 150 + 240 + 180 = 645
Link flow 4-5: 150 + 200 + 350 + 250 + 50 + 100 + 125 + 225 = 1450
Link flow 5-6: 125 + 225 + 150 + 150 + 200 + 175 = 1025

Figure 2.2 illustrates these 2-way daily link volumes.

Contd

Example 2.7 Contd

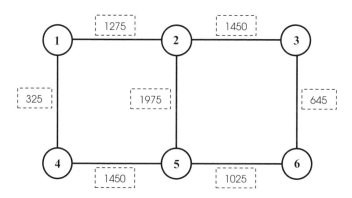

Figure 2.2 Link volumes arising from 'all-or-nothing' traffic assignment procedure.

Table 2.14 Minimum time/cost paths between zones in transport network

Origin zone	Destination zone					
	1	2	3	4	5	6
1		1-2	1-2-3	1-4	1-2-5	1-2-3-6
2	2-1		2-3	2-5-4	2-5	2-5-6
3	3-2-1	3-2		3-2-5-4	3-2-5	3-6
4	4-1	4-5-2	4-5-2-3		4-5	4-5-6
5	5-2-1	5-2	5-2-3	5-4		5-6
6	6-3-2-1	6-5-2	6-3	6-5-4	6-5	

Table 2.15 Trip interchanges between the six zones

Origin zone	Destination zone					
	1	2	3	4	5	6
1		250	100	125	150	75
2	300		275	200	400	150
3	150	325		100	100	240
4	200	150	50		350	125
5	100	300	125	250		200
6	150	150	180	225	175	

2.8   A full example of the four-stage transportation modelling process

2.8.1   *Trip production*

Assume a study area is divided into seven zones (A, B, C, D, E, F, G) as indicated in Fig. 2.3. Transport planners wish to estimate the volume of car traffic for each of the links within the network for ten years into the future (termed the design year).

Using land use data compiled from the baseline year on the trips attracted to and generated by each zone, together with information on the three main trip generation factors for each of the seven zones:

- Population (trip productions)
- Retail floor area (trip attractions)
- Employment levels (trip attractions)

linear regression analysis yields the following zone-based equations for the two relevant dependent variables (zonal trip productions and zonal trip attractions) as follows:

$$P = (3 \times \text{population}) - 500 \tag{2.17}$$
$$A = (3 \times \text{number employed}) + (75 \times \text{office floor space, m}^2) + 400 \tag{2.18}$$

Table 2.16 gives zonal trip generation factors for the design year, together with the trip productions and attractions estimated from these factors using Equations 2.17 and 2.18.

Table 2.16 Trip productions and attractions for the design year (10 years after baseline year)

Zone	Population	Office floor area (m²)	Numbers employed	Trip productions	Trip attractions
A	7500	50	775	22000	6475
B	4000	400	3500	11500	40900
C	6000	75	700	17500	8125
D	5000	250	4000	14500	31150
E	9000	100	1000	26500	10900
F	6000	50	3000	17500	13150
G	4000	100	800	11500	10300
Total	41500	1025	13775	121000	121000

For example, in the case of zone A:

Trips produced $= 3 \times 7500 - 500 = 22000$

Trips attracted $= (3 \times 775) + (75 \times 50) + 400 = 6475$

2.8.2   Trip distribution

In order to compile the trip distribution matrix, the impedance term relating to the resistance to travel between each pair of zones must be established. In this case, the travel time is taken as a measure of the impedance and the zone-to-zone times are given in Table 2.17.

Using a gravity model with the deterrence function in the following form between zone i and zone j:

$$F_{ij} = t_{ij}^{-2}$$

where t_{ij} is the time taken to travel between zone i and zone j

The interzonal trips are estimated using Equation 2.3. For example, taking the trips from zone A to all other zones, it can be seen from Table 2.16 that 6475

Table 2.17 Interzonal travel times

Origin zone	A	B	C	D	E	F	G
				Destination zone			
A		10	15	15	20	25	32
B	10		7	5	10	15	22
C	15	7		8	14	16	26
D	15	5	8		6	10	18
E	20	10	14	6		16	12
F	25	15	16	10	16		12
G	32	22	26	18	12	12	

Table 2.18 Gravity model computations for Zone A

Zone	A_j	P_i	T_{ij}	F_{ij}	$P_i \times F_{ij}$	$\dfrac{P_iF_{ij}}{\sum_j(P_iF_{ij})}$	$\dfrac{A_jP_iF_{ij}}{\sum_j(P_iF_{ij})}$
A	6475	22 000					
B to A		11 500	10	0.010	115.0	0.317	2053.0
C to A		17 500	15	0.004	77.78	0.214	1388.5
D to A		14 500	15	0.004	64.44	0.178	1150.5
E to A		26 500	20	0.003	66.25	0.183	1182.7
F to A		17 500	25	0.002	28.00	0.077	499.80
G to A		11 500	32	0.001	11.23	0.031	200.50
					$\Sigma = 362.7$	$\Sigma = 1$	$\Sigma = 6475$

trips were attracted to zone A. Equation 2.3 is used to estimate what proportion of this total amount sets out from each of the other six zones, based on the relative number of trips produced by each of the six zones and the time taken to travel from each to zone A. These computations are given in Table 2.18.

When an identical set of calculations are done for the other six zones using the gravity model, the initial trip matrix shown in Table 2.19 is obtained.

It can be seen from Table 2.19 that, while each individual column sums to give the correct number of trips attracted for each of the seven zones, each indi-

Table 2.19 Initial output from gravity model

Origin zone	A	B	C	D	E	F	G	Total
				Destination zone				
A		5905	1019	1713	740	958	525	10 861
B	2053		2446	8060	1547	1391	581	16 078
C	1388	9587		4791	1201	1861	633	19 461
D	1150	15 569	2361		5418	3947	1094	29 540
E	1183	7113	1409	12 898		2818	4498	29 919
F	500	2088	712	3066	920		2970	10 256
G	200	638	177	622	1074	2174		4886
Total	6475	40 900	8125	31 150	10 900	13 150	10 300	121 000

vidual row does not sum to give the correct number of trips produced by each. (It should be noted that the overall number of productions and attractions are equal at the correct value of 121 000.)

In order to produce a final matrix where both rows and columns sum to their correct values, a remedial procedure must be undertaken, termed the row-column factor technique. It is a two-step process.

First, each row sum is corrected by a factor that gives the zone in question its correct sum total (Table 2.20).

Second, because the column sums no longer give their correct summation, these are now multiplied by a factor which returns them to their correct individual totals (Table 2.21).

This repetitive process is continued until a final matrix is obtained where the production and attraction value for each zone is very close to the correct row and column totals (Table 2.22).

Table 2.20 Row correction of initial gravity model trip matrix

Origin zone	Destination zone							Total	Correct total	Row factor
	A	B	C	D	E	F	G			
A		5905	1019	1713	740	958	525	10861	22000	2.026
B	2053		2446	8060	1547	1391	581	16078	11500	0.715
C	1388	9587		4791	1201	1861	633	19461	17500	0.899
D	1150	15569	2361		5418	3947	1094	29540	14500	0.491
E	1183	7113	1409	12898		2818	4498	29919	26500	0.886
F	500	2088	712	3066	920		2970	10256	17500	1.706
G	200	638	177	622	1074	2174		4886	11500	2.354
Total	6475	40900	8125	31150	10900	13150	10300	121000	121000	

Table 2.21 Column correction of gravity model trip matrix

Origin zone	Destination zone							Total
	A	B	C	D	E	F	G	
A	0	11962	2064	3470	1499	1941	1064	22000
B	1468	0	1750	5765	1107	995	415	11500
C	1249	8621	0	4308	1080	1673	569	17500
D	565	7642	1159	0	2660	1938	537	14500
E	1048	6301	1248	11424	0	2496	3984	26500
F	853	3562	1216	5232	1569	0	5068	17500
G	472	1501	417	1464	2529	5117	0	11500
Total	5654	39589	7854	31663	10443	14160	11636	
Correct total	6475	40900	8125	31150	10900	13150	10300	121000
Column factor	1.145	1.033	1.035	0.984	1.044	0.929	0.885	

Table 2.22 Final corrected trip matrix

Origin zone	Destination zone							Total
	A	B	C	D	E	F	G	
A		12 286	2112	3352	1 551	1 780	918	22 000
B	1670	0	1800	5 599	1 152	918	361	11 500
C	1407	8 818	0	4 144	1 114	1 528	489	17 500
D	632	7 759	1172	0	2 722	1 757	458	14 500
E	1222	6 673	1317	11 380	0	2 360	3 548	26 500
F	998	3 784	1286	5 226	1 680	0	4 526	17 500
G	547	1 579	437	1 448	2 681	4 807	0	11 500
Total	6475	40 900	8125	31 150	10 900	13 150	10 300	121 000

(Note, if Equation 2.2 is used within the trip distribution process, the rows sum correctly whereas the columns do not. In this situation the row-column factor method is again used but the two-stage process is reversed as a correction is first applied to the column totals and then to the new row totals.)

2.8.3   Modal split

Two modes of travel are available to all trip makers within the interchange matrix: bus and private car. In order to determine the proportion of trips undertaken by car, the utility of each mode must be estimated. The utility functions for the two modes are:

$$U_{CAR} = 2.5 - 0.6C - 0.01T \qquad (2.19)$$

$$U_{BUS} = 0.0 - 0.6C - 0.01T \qquad (2.20)$$

where
C = cost (£)
T = travel time (minutes)

For all travellers between each pair of zones:

- The trip by car costs £2.00 more than by bus
- The journey takes 10 minutes longer by bus than by car.

Since the model parameters for the cost and time variables are the same in Equations 2.19 and 2.20, the relative utilities of the two modes can be easily calculated:

$$\begin{aligned} U_{(BUS-CAR)} &= (0.0 - 2.5) - 0.6(c - (c + 2)) - 0.01((t + 10) - t) \\ &= -2.5 + 1.2 - 0.1 \\ &= -1.4 \end{aligned}$$

$$U_{(CAR-BUS)} = (2.5 - 0.0) - 0.6((c + 2) - c) - 0.01(t - (t + 10))$$
$$= 2.5 - 1.2 + 0.1$$
$$= 1.4$$

where
£c = cost of travel by bus
£(c + 2) = cost of travel by car
t = travel time by car (in minutes)
(t + 10) = travel time by bus (in minutes)

We can now calculate the probability of the journey being made by car using Equation 2.16:

$$P_{BUS} = 1 \div \left(1 + e^{(U_{CAR} - U_{BUS})}\right)$$
$$= 1 \div \left(1 + e^{(1.4)}\right)$$
$$= 0.198$$
$$P_{CAR} = 1 \div \left(1 + e^{(U_{BUS} - U_{CAR})}\right)$$
$$= 1 \div \left(1 + e^{(-1.4)}\right)$$
$$= 0.802$$

So just over 80% of all trips made will be by car. If we assume that each car has, on average, 1.2 occupants, multiplying each cell within Table 2.22 by 0.802 and dividing by 1.2 will deliver a final matrix of car trips between the seven zones as shown in Table 2.23.

Table 2.23 Interzonal trips by car

Origin zone	A	B	C	D	E	F	G
A	0	8213	1412	2241	1037	1190	614
B	1117	0	1203	3743	770	613	241
C	940	5895	0	2771	744	1022	327
D	422	5187	784	0	1820	1174	306
E	817	4461	880	7607	0	1578	2372
F	667	2529	860	3494	1123	0	3025
G	366	1056	292	968	1792	3213	0

Destination zone (column header spanning A–G)

2.8.4   Trip assignment

The final stage involves assigning all the car trips in the matrix within Table 2.23 to the various links within the highway network shown in Fig. 2.3. Taking the information on the interzonal travel times in Table 2.17 and using the 'all-or-nothing' method of traffic assignment, the zone pairs contributing to the flow along each link can be established (Table 2.24). The addition of the flows from each pair along a given link allows its 2-way flow to be estimated. These are shown in Fig. 2.4.

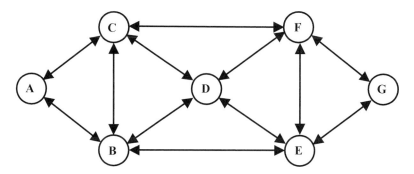

Figure 2.3 Zones and links in study area within worked example.

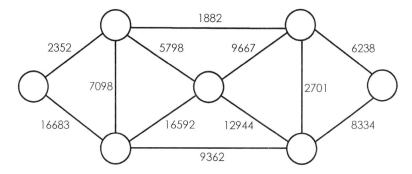

Figure 2.4 Interzonal link flows for private vehicles (cars).

Table 2.24 2-way vehicular flows along each link

Network link	Zone pairs contributing to flow along link	Total link flow
A to B	(A,B)(B,A) (A,D)(D,A) (A,E)(E,A) (A,F)(F,A) (A,G)(G,A)	16 683
A to C	(A,C)(C,A)	2 352
B to C	(B,C)(C,B)	7 098
B to D	(A,D)(D,A) (A,F)(F,A) (B,D)(D,B) (B,F)(F,B)	16 592
B to E	(A,E)(E,A) (A,G)(G,A) (B,E)(E,B) (B,G)(G,B)	9 362
C to D	(C,D)(D,C) (C,E)(E,C) (C,G)(G,C)	5 798
C to F	(C,F)(F,C)	1 882
D to E	(C,E)(E,C) (C,G)(G,C) (D,E)(E,D) (D,G)(G,D)	12 944
D to F	(A,F)(F,A) (B,F)(F,B) (D,F)(F,D)	9 667
E to F	(E,F)(F,E)	2 701
E to G	(A,G)(G,A) (B,G)(G,B) (C,G)(G,C) (D,G)(G,D) (E,G)(G,E)	8 334
F to G	(F,G)(G,F)	6 238

2.9   Concluding comments

The process of traffic forecasting lies at the very basis of highway engineering. Modelling transport demand is normally undertaken using a four-stage sequential process starting with trip generation and distribution, followed by modal split and concluding with traffic assignment. Predicting flows along the links within a highway network provides vital information for the economic and environmental assessments required as part of the project appraisal process and allows the scale of each individual project within the network to be determined. Once the demand analysis and appraisal process have been completed, the detailed junction and link design can then be undertaken.

It should be remembered, however, that the modelling process is a simplification of reality. Predictions arising from it are broad estimates rather than precise forecasts. The error range within which the model results are likely to fall should accompany any data supplied to the transport planners.

2.10   References

Furness, K.P. (1965) Time function iteration. *Traffic Engineering Control*, 7, 458–460.

McFadden, D. (1981) Economic models and probabilistic choice. In *Structural Analysis of Discrete Data with Econometric Applications* (eds Manski & McFadden). MIT Press, Cambridge, MA, USA.

Wardrop, J.G. (1952) Some theoretical aspects of road traffic research. *Proceedings of the Institution of Civil Engineers*, **1**(36), 325–362.

Chapter 3
Scheme Appraisal for Highway Projects

3.1 Introduction

Once a transportation plan has been finalised and the demand along each of its highway links has been established, a process must be put in place that helps identify the best solution for each individual proposal within the highway network. Each project must therefore be subject to an appraisal.

The aim of the highway appraisal process is therefore to determine the economic, societal and environmental feasibility of the project or group of projects under examination. The process enables highway planners to decide whether a project is desirable in absolute terms and also provides a means of choosing between different competing project options, all of which have the ability to meet the stated goals and objectives of the project sponsors.

The 'reasoned choice' model of individual or group decisions provides a decision-making framework within which scheme appraisal can take place, providing a technical foundation for non-recurring decisions such as the assessment of a highway construction/improvement proposal (Zey, 1992). It comprises the following steps:

(1) *Problem recognition.* The decision-maker determines that a problem exists and that a decision must be reflected on.
(2) *Goal identification.* The decision-maker details the desired result or outcome of the process.
(3) *Identification of alternative highway schemes.* Different potential solutions are assembled prior to their evaluation.
(4) *Information search.* The decision-maker seeks to identify characteristics associated with the alternative solutions.
(5) *Assessment of information on alternative highway schemes.* The information necessary for making a decision regarding the preferred option is gathered together and considered.
(6) *Selection of preferred highway scheme.* A preferred option is selected by the decision-maker for implementation in the future.
(7) *Evaluation.* The decision is assessed a period of time after its implementation in order to evaluate it on the basis of its achieved results.

Clear rationality, where a judgement is arrived at following a sequence of deliberately followed logical steps, lies at the basis of this model for decision-making. The principles of reasoned choice have been adapted into an analytic technique, called the rational approach, which has been detailed in Chapter 1.

The scheme appraisal process for highway schemes can be broken down broadly into two sections: economic evaluation and environmental assessment. Background details of these two types of assessment have been given in Chapter 1. Each of these is addressed in some detail below, and this chapter also deals with an appraisal technique introduced in the UK that combines these two types of highway project evaluation.

3.2 Economic appraisal of highway schemes

At various points in the development of a highway project, the developer will require economic assessments of the route options under consideration. This will involve comparing their performance against the current situation, termed the 'do-nothing' alternative, and/or against the 'do-minimum' alternative involving a low-cost upgrading of the existing facility. Computations are performed on the costs and benefits associated with each highway option in order to obtain one or more measures of worth for each. Engineering economics provides a number of techniques that result in numerical values termed measures of economic worth. These, by definition, consider the time value of money, an important concept in engineering economics that estimates the change in worth of an amount of money over a given period of time. Some common measures of worth are:

- Net present value (NPV)
- Benefit/cost ratio (B/C)
- Internal rate of return (IRR).

In economic analysis, financial units (pounds/euros/dollars) are used as the tangible basis of evaluation. With each of the above 'measure of worth' techniques, the fact that a quantity of money today is worth a different amount in the future is central to the evaluation.

Within the process of actual selection of the best option in economic terms, some criterion based on one of the above measures of worth is used to select the chosen proposal. When several ways exist to accomplish a given objective, the option with the lowest overall cost or highest overall net income is chosen. While intangible factors that cannot be expressed in monetary terms do play a part in an economic analysis, their role in the evaluation is, to a large extent, a secondary one. If, however, the options available have approximately the same equivalent cost/value, the non-economic and intangible factors may be used to select the best option.

Economic appraisal techniques can be used to justify a scheme in absolute terms, in which case the decision is made on the basis of whether the project is 'economically efficient' or not. A negative net present value or a benefit/cost ratio less than unity would indicate an inefficient scheme where society would end up worse off with the scheme than without it. The economic benefits accruing to the beneficiaries of the highway would be exceeded by economic costs incurred by those 'losing out' as a result of its construction. In the main, the beneficiaries are the road users and the 'losers' are those funding the scheme. Where the appraisal is being used to help differentiate between the economic performances of competing options under examination, the scheme with the highest measure of worth will be deemed the most efficient, assuming that at least one will have a positive NPV or a B/C ratio greater than unity.

The framework within which this evaluation of the economic consequences of highway schemes takes place is referred to as cost-benefit analysis.

3.3   Cost-benefit analysis (CBA)

3.3.1   Introduction

Within Europe, the method usually adopted for the economic evaluation of highway schemes, termed cost-benefit analysis, utilises the net present value technique where the costs and benefits of the scheme are discounted over time so that they represent present day values. Using this method, any proposal having a positive net present value is economically sustainable in absolute terms. Where competing project options are being compared, assuming they are being used in identical capacities over the same period, the one with the numerically larger NPV is selected (i.e. the one that is less negative or more positive). A brief historical background to the method has already been given in Chapter 1. The main steps in the technique involve the listing of the main project options, the identification and discounting to their present values of all relevant costs and benefits required to assess them, and the use of economic indicators to enable a decision to be reached regarding the proposal's relative or absolute desirability in economic terms.

3.3.2   Identifying the main project options

This is a fundamental step in the CBA process where the decision-makers compile a list of all relevant *feasible* options that they wish to be assessed. It is usual to include a 'do-nothing' option within the analysis in order to gauge those evaluated against the baseline scenario where no work is carried out. The 'do-minimum' option offers a more realistic course of action where no new highway is constructed but a set of traffic management improvements are made to the existing route in order to improve the overall traffic performance. Evaluation of the 'do-nothing' scenario does however ensure that, in addition to the various 'live' options being compared in relative terms, these are also seen to be economically justified in absolute terms, in other words their benefits exceed their costs.

The term 'feasible' refers to options that, on a preliminary evaluation, present themselves as viable courses of action that can be brought to completion given the constraints imposed on the decision-maker such as lack of time, information and resources.

Finding sound feasible options is an important component of the decision process. The quality of the final outcome can never exceed that allowed by the best option examined. There are many procedures for both identifying and defining project options. These include:

- Drawing on the personal experience of the decision-maker himself as well as other experts in the highway engineering field
- Making comparisons between the current decision problem and ones previously solved in a successful manner
- Examining all relevant literature.

Some form of group brainstorming session can be quite effective in bringing viable options to light. Brainstorming consists of two main phases. Within the first, a group of people put forward, in a relaxed environment, as many ideas as possible relevant to the problem being considered. The main rule for this phase is that members of the group should avoid being critical of their own ideas or those of others, no matter how far-fetched. This non-critical phase is very difficult for engineers, given that they are trained to think analytically or in a judgmental mode (Martin, 1993). Success in this phase requires the engineer's judgmental mode to be 'shut down'. This phase, if properly done, will result in the emergence of a large number of widely differing options.

The second phase requires the planning engineer to return to normal judgmental mode to select the best options from the total list, analysing each for technological, environmental and economic practicality. This is, in effect, a screening process which filters through the best options. One such method is to compare each new option with an existing, 'tried-and-tested' option used in previous similar highway proposals by means of a T-chart (Riggs *et al.*, 1997). The chart contains a list of criteria which any acceptable option should satisfy. The option under examination is judged on the basis of whether it performs better or worse than the conventional option on each of the listed criteria. It is vital that this process is undertaken by highway engineers with the appropriate level of experience, professional training and local knowledge in order that a sufficiently wide range of options arise for consideration.

An example of a T-chart is shown in Table 3.1.

In the example in Table 3.1, the proposed option would be rejected on the basis that, while it had a lower construction cost, its maintenance costs and level of environmental intrusion and geometric design, together with its low level of

Proposed highway option vs. an accepted 'tried and tested' design solution	Better	Worse
Construction cost	✓	
Maintenance cost		✓
Environmental impact		✓
Geometric design		✓
Time savings		✓

Table 3.1 Example of T-chart for a highway project

time savings for motorists, would eliminate it from further consideration. The example illustrates a very preliminary screening process. A more detailed, finer process would involve the use of percentages rather than checkmarks. The level of filtering required will depend on the final number of project options you wish to be brought forward to the full evaluation stage.

3.3.3   Identifying all relevant costs and benefits

The application of cost-benefit for project assessment in the highway area is made more complicated by the wide array of benefits associated with a given road initiative, some easier to translate into monetary values than others. Many of the benefits of improvements to transport projects equate to decreases in cost. The primary grouping that contains this type of economic gain is termed user benefits. Benefits of this type accrue to those who will actively use the proposed installation. This grouping includes:

- Reductions in vehicle operating costs
- Savings in time
- Reduction in the frequency of accidents.

This is the main group of impacts considered within a standard highway CBA. Other studies might address in some way a secondary grouping of benefits – those accruing to 'non-users' of the proposed facility. These include:

- Positive or negative changes in the environment felt by those people situated either near the new route or the existing route from which the new one will divert traffic. These can be measured in terms of the changes in impacts such as air pollution, noise or visual intrusion/obstruction.
- The loss or improvement of recreational facilities used by local inhabitants, or the improvement or deterioration in access to these facilities.

The costs associated with a proposed highway installation can fall into similar categories. However, in most evaluations, construction costs incurred during the initial building phase, followed by maintenance costs incurred on an ongoing basis throughout the life of the project, are sufficient to consider.

The three primary user benefits listed above are normally estimated relative to the *without project* or 'do-nothing' situation. The definition and description of the without project scenario should be such that it constitutes an entirely feasible and credible course of action. Let us examine each of these benefits in some detail.

Reductions in vehicle operating costs

This constitutes the most direct potential benefit derived from a new or upgraded highway project. It is often the most important one and the one easiest to measure in money terms. While the users are the initial beneficiaries of these potential reductions, circumstances dictated by government policies or competition, or the drive to maximise profits, might lead to other groups within the broader community having a share in the ultimate benefit.

For a highway scheme, the new upgraded project leads to lower levels of congestion and higher speeds than on the existing roadway, usually resulting in lower fuel consumption and lower maintenance costs due to the reduced wear and tear on the vehicles.

Within a highway cost-benefit analysis, a formula is used which directly relates vehicle-operating costs to speed. Costs included are both fuel and non-fuel-based. The higher speeds possible on the new road relative to the existing one lead to potential monetary savings for each road user.

Savings in time

The upgrading of a highway installation will invariably reduce travel time as well as improving the reliability of transport services. For transport users, time has some connection with money. The degree of correlation between the two depends primarily on the manner in which the opportunities made possible by the increased availability of time are utilised.

In general, analyses of the value of time-savings within the cost-benefit framework focus on distinguishing between travel for work and travel for non-work purposes. Non-work time includes leisure travel and travel to and from work. Within developed economies, the value of working time is related to the average industrial wage plus added fringe benefits, on the assumption that time saved will be diverted to other productive uses. There is no broad agreement among economic evaluation experts regarding the valuation of non-work time. Since there is no direct market available that might provide the appropriate value, values must be deduced from the choices members of the public make that involve differences in time. Studies carried out in industrialised countries have indicated that travellers value non-working time at between 20% and 35% the value attributed to working time (Adler, 1987). Less developed countries may, however, set the valuation at a lower percentage. In the worked example presented in section 3.3.6, an average value for time savings is used which supplies a single value covering both workers and non-workers using the highway.

Reduction in the frequency of accidents

Assessing the economic benefit of accident reduction entails two steps. In the case of a highway, this requires comparison of the accident rate on the existing unimproved highway with that of other highways elsewhere in the country (or abroad) constructed to the higher standard of the proposed new road. Normally, the higher the standard of construction of a highway, the lower its accident rate.

The second step involves the monetary valuation of the accident reduction. Three types of damage should be considered:

- Property damage
- Personal injuries arising from serious accidents
- Fatal accidents.

Property damage to vehicles involved in accidents is the most easily measured in money terms. Reduced breakage of cargo can also be a significant benefit in proposed rail-based and seaport installations. Valuations can be obtained directly from the extent of claims on insurance policies.

The cost of serious but non-fatal accidents is much more difficult to assess. Medical costs and the cost of lost output and personal pain and suffering constitute a large proportion of the total valuation.

There is major disagreement on which method is most appropriate for estimating the economic cost to society of a fatal accident. In recent times, stated preference survey techniques have been employed to estimate this valuation.

In most cases, an average cost per accident, covering fatal and non-fatal, is employed, with damage costs also accounted for within the final estimated value.

3.3.4   Economic life, residual value and the discount rate

A highway project is often complex and long term, with the costs and benefits associated with it occurring over a long time frame which we term the *life* of the project, a parameter dealt with in earlier chapters. It sets a limit on the period over which the costs and benefits are estimated, as all must occur within this time slot, be it 25, 35 or even 50 years or more. It is related, in principle, to the expected lifetime of the project under analysis.

Given that transport development projects have the potential to be in service for a very long time, it may seem impossible to set a limit on the life of the project with any degree of certainty. In practice, however, this may not give rise to serious problems in the evaluation, as the loss of accuracy that results from limiting the life of a project to 35–40 years, instead of continuing the computation far beyond this point, is marginal to the analyst undertaking the evaluation. The shortened analysis can be justified on the basis that, in time equivalent terms, substantial costs and/or benefits are unlikely to arise in the latter years of the project. If they are predicted, the life may well have to be extended. Truncating the analysis can also be justified on the basis of the uncertainty with which costs and benefits that occur beyond a certain time horizon can be predicted.

Where this technique is applied after a relatively small number of years, the project may well have to be assigned a substantial residual or salvage value, reflecting the significant benefits still to be accrued from the project or, conversely, costs still liable to be incurred by it (a residual value can be negative, as say for a nuclear power station yet to be decommissioned). The difficulty in assigning a meaningful residual value to a project after so few years in commission results in this solution being rather unsatisfactory. It is far more advisable to extend the evaluation to a future point in time where the residual value is extremely small relative to its initial value.

In addition to this, the costs and benefits occur at different times over this time horizon. Because of this, they cannot be directly combined until they are reduced to a common time frame. This is achieved using another parameter introduced earlier, the *discount rate*, which translates all costs and benefits to time equivalent values. The actual value used is the social discount rate, given that the decision-maker is interested in the benefits and costs to society as a whole rather than to any individual or group of individuals.

The setting of this rate is quite a complex process, and is somewhat beyond the scope of this text. It is important to point out, however, that it is not the same as the market interest rate available to all private borrowers. It is a collective discount rate reflecting a project of benefit to a large number of people and spanning a time frame greater than one full generation. A single definitive discount rate does not exist. Its estimation can be based on time preference or the opportunity cost of resources. The first is based on people in general having a preference for development taking place now rather than in the future. Because this involves taking a long-term view, the social time preference rate is usually set at a low, single-figure rate. The second reflects what members of society have foregone as a result of funds being devoted to the development in question. The prevailing real interest rate is often used as a guide for this value. Typical rates can reach 15%, appreciably higher than the figure obtained from the time preference approach. Economists will have varying views about the most appropriate test discount rate to use. In many instances the main decision-maker or the person financing the proposal will set the rate. Before doing so, discussions with all relevant stakeholders may be appropriate.

3.3.5   *Use of economic indicators to assess basic economic viability*

Once the two parameters of project life and discount rate are set in place, these allow all costs and benefits to be directly compared at the same point in time. The decision-maker must now choose the actual mechanism for comparing and analysing the costs and benefits in order to arrive at a final answer for the net benefit of each of the project options under consideration. Any of the three techniques listed earlier in the chapter can be used for this purpose:

● Net present value (NPV)
● Internal rate of return (IRR)
● Benefit/cost ratio (B/C).

The NPV will estimate the economic worth of the project in terms of the present worth of the total net benefits. The IRR will give, for each option under consideration, the rate at which the net present value for it equals zero, with the B/C ratio based on the ratio of the present value of the benefits to the present value of the costs. For the last two methods, if the options under consideration are mutually exclusive, an incremental analysis must be carried out to establish the best performing one in economic terms.

All three methods depend on discounting to arrive at a final answer. All, if used correctly, should give answers entirely consistent with each other, but the specific technique to be used varies with the circumstances. Thus, while the chosen technique is, to a certain extent, down to the preference of the decision-maker, it is nonetheless dependent on the type of decision to be taken within the analysis. If the decision is whether or not to proceed with a given project, the result from the chosen technique is compared with some predetermined threshold value in order to decide whether the project is economically justified. Once a discount rate/minimum acceptable rate of return is set, any of the above methods will give the same result. Assuming a discount rate of 10%, the project will be economically acceptable if the NPV of the net benefits at 10% exceeds zero, if the IRR is above 10% or if the B/C ratio at 10% exceeds unity.

In the case of an independent project where choosing one does not exclude the possibility of proceeding with one or all of the others, all techniques yield the same result, the critical question being the choice of discount rate. In choosing between mutually exclusive projects where choice of one immediately excludes all others, the most straightforward method involves choosing the option with the maximum NPV of net benefits.

There may however be situations where it is required to rank order a number of highway projects, on the basis that there is a set quantity of resources available for developing a certain category of project, and the decision-maker wishes to have a sequence in which these projects should be approved and constructed until the allotted resources are exhausted. In these cases, ranking based on NPV may be of limited assistance, since high cost projects with slightly greater NPV scores may be given priority over lower cost ones yielding greater benefits per unit cost. A correct course of action would be to rank the different project options based on their benefit/cost ratio, with the one with the highest

B/C score given the rank 1, the second highest score given the rank 2, and so on.

Selecting a criterion for deciding between project options can be contentious. Some decision-makers are used to incorporating certain techniques in their analyses and are loath to change. IRR is rarely mentioned in the preceding paragraph, yet a number of national governments have a preference for it. This inclination towards it by some decision-makers is to some extent based on the fact that many have a background in banking and thus have an innate familiarity with this criterion, together with the perception that its use does not require a discount rate to be assumed or agreed. The latter statement is, in fact, incorrect, as, particularly when evaluating a single project, IRR must be compared with some agreed discount rate.

Other supplementary methods of analysis such as cost effectiveness analysis and the payback period could also be used to analyse project options. Details of the payback method are given later in this chapter.

3.3.6 Highway CBA worked example

Introduction

It is proposed to upgrade an existing single carriageway road to a dual carriageway and to improve some of the junctions. The time frame for construction of the scheme is set at two years, with the benefits of the scheme accruing to the road users at the start of the third year. As listed above, the three main benefits are taken as time savings, accident cost savings and vehicle operating cost reductions. Construction costs are incurred mainly during the two years of construction, but ongoing annual maintenance costs must be allowed for throughout the economic life of the project, taken, in this case, to be 10 years after the road has been commissioned.

The following basic data is assumed for this analysis:

Accident rates:	0.85 per million vehicle-kilometres (existing road)
	0.25 per million vehicle-kilometres (upgraded road)
Average accident cost:	£10 000
Average vehicle time savings:	£2.00 per hour
Average vehicle speeds:	40 km/h (existing road)
	85 km/h (upgraded road)
Average vehicle operating cost:	$((2 + (35/V) + 0.00005*V^2) \div 100)$ £ per km
Discount rate:	6%

The traffic flows and the construction/maintenance costs for the highway proposal are shown in Table 3.2.

Year	Predicted flow (10⁶ v.km/yr)	Construction cost (£)	Operating cost (£)
1	—	15 000 000	—
2	—	10 000 000	—
3	250	—	500 000
4	260	—	500 000
5	270	—	500 000
6	280	—	500 000
7	290	—	500 000
8	300	—	500 000
9	310	—	500 000
10	320	—	500 000
11	330	—	500 000
12	340	—	500 000

Table 3.2 Traffic flows and costs throughout economic life of the highway proposal

Computation of discounted benefits and costs

Table 3.3 gives the valuations for the three user benefits over the 10 years of the upgraded highways operating life.

Table 3.3 Valuations of discounted highway user benefits

Year	Accident cost savings (£)	Operating cost savings (£)	Travel time savings (£)	Total user benefits (£)	Discounted benefits (£)
1	—	—	—	—	—
2	—	—	—	—	—
3	1 500 000	454 963	6 617 647	8 572 610	7 197 729
4	1 560 000	473 162	6 882 353	8 915 515	7 061 923
5	1 620 000	491 360	7 147 059	9 258 419	6 918 429
6	1 680 000	509 559	7 411 765	9 601 324	6 768 554
7	1 740 000	527 757	7 676 471	9 944 228	6 613 480
8	1 800 000	545 956	7 941 176	10 287 132	6 454 274
9	1 860 000	564 155	8 205 882	10 630 037	6 291 902
10	1 920 000	582 353	8 470 588	10 972 941	6 127 233
11	1 980 000	600 552	8 735 294	11 315 846	5 961 046
12	2 040 000	618 750	9 000 000	11 658 750	5 794 042
					$\Sigma = 65 188 612$

Taking the computations for year 7 as an example, the three individual user benefits together with their total and discounted value are calculated as follows:

Accident savings $_{(Yr\ 7)}$
$$= (0.85 - 0.25) \times 10 000 \times 290$$
$$= £1 740 000$$

Operating cost $_{(existing\ route)}$
$$= (2 + 35/40 + (0.00005 \times 40^2)) \div 100$$
$$= £0.02955 \text{ per km per vehicle}$$

Operating cost $_{(upgraded\ route)}$
$$= (2 + 35/85 + (0.00005 \times 85^2)) \div 100$$
$$= £0.02773 \text{ per km per vehicle}$$

Operation savings $_{(Yr\ 7)}$	$= (0.02955 - 0.02773) \times 290 \times 10^6$
	$= £527\,757$
Travel time/km $_{(existing\ route)}$	$= 1/40$
	$= 0.025$ hours
Travel time/km $_{(upgraded\ route)}$	$= 1/85$
	$= 0.011765$ hours
Value of savings per veh-km	$= (0.025 - 0.0117647) \times £2.00$
	$= £0.02647$
Value of time savings $_{(Yr\ 7)}$	$= 0.02647 \times 290 \times 10^6$
	$= £7\,676\,471$
Total benefits $_{(Yr\ 7)}$	$= (1\,740\,000 + 527\,757 + 7\,676\,471)$
	$= £9\,944\,228$
Discounted benefits $_{(Yr\ 7)}$	$= 9\,944\,229 \div (1.06)^7$
	$= 9\,944\,229 \div 1.50363$
	$= £6\,613\,480$

These calculated figures are given in row seven of Table 3.3. The results of the computation of user benefits for all relevant years within the highway's economic life are shown in this table. It can be seen that the discounted value of the total benefits amounts to £65 188 612.

Table 3.4 gives the construction and maintenance costs incurred by the project over its economic life together with the discounted value of these costs.

As seen from Table 3.4, the total value of the discounted costs of the upgrading project is estimated at £26 326 133.

The computations contained in Tables 3.3 and 3.4 are used to estimate the economic worth of the project. This can be done using the three indicators referred to earlier in the chapter: net present value, benefit/cost ratio and internal rate of return.

Table 3.4 Valuation of discounted construction/maintenance costs

Year	Construction and maintenance costs (£)	Discounted costs (£)
1	15 000 000	14 150 943
2	10 000 000	8 899 964.4
3	500 000	419 809.64
4	500 000	396 046.83
5	500 000	373 629.09
6	500 000	352 480.27
7	500 000	332 528.56
8	500 000	313 706.19
9	500 000	295 949.23
10	500 000	279 197.39
11	500 000	263 393.76
12	500 000	248 484.68
		$\Sigma = 26\,326\,133$

Net present value

To obtain this figure, the discounted costs are subtracted from the discounted benefits:

$$NPV = 65\,188\,612 - 26\,326\,133$$
$$= £38\,862\,479$$

Benefit-cost ratio

In this case, the discounted benefits are divided by the discounted costs as follows:

$$B/C \text{ ratio} = 65\,188\,612 \div 26\,326\,133$$
$$= 2.476$$

Internal rate of return

This measure of economic viability is estimated by finding the discount rate at which the discounted benefits equate with the discounted costs. In this example, this occurs at a rate of 28.1%.

Summary

All the above indicators point to the economic strength of the project under examination. Its NPV at just over £38 million is strongly positive, and its B/C ratio at just below 2.5 is well in excess of unity. The IRR value of over 28% is over four times the agreed discount rate (6%). Together they give strong economic justification for the project under examination. Knowledge of these indicators for a list of potential projects will allow decision-makers to compare them in economic terms and to fast track those that deliver the maximum net economic benefit to the community.

3.3.7   COBA

Within the formal highway appraisal process in the UK for trunk roads, cost-benefit analysis is formally carried out using the COBA computer program (DoT, 1982) which assesses user costs and benefits over a 30-year period – assumed to be the useful life of the scheme – in order to obtain its net present value. (The current version of the programme is COBA 9.) This is divided by the initial capital cost of the scheme and expressed in percentage terms to give the COBA rate of return.

The COBA framework involves comparing each alternative proposal with the 'do-minimum' option, with the resulting net costs and benefits providing the input to COBA. For example, if a choice is required between route A, route B and neither, then the costs and benefits of neither would be subtracted from each of the A and B valuations before the cost-benefit computation is made.

The output from COBA is used to contribute to the following type of decision:

- To assess the need for improving a specific highway route. The improvement could involve either the upgrading of an existing roadway or the construction of a completely new one
- To determine what level of priority should be assigned to a particular scheme by considering its economic return relative to those of the other viable schemes in the area/region being considered by road administrators
- To determine the optimal timing of the scheme in question relative to other road schemes in the area
- To aid in the presentation of viable highway options to the public within a formal consultation process
- To establish optimal junction and link designs by comparing the economic performance of the options under consideration.

The extent to which a full COBA analysis can be undertaken for a particular scheme depends on the stage reached in the assessment process, the data available to the decision-maker and the nature of the decision to be taken. As the design procedure for a particular scheme advances, a more refined economic analysis becomes possible.

Within COBA, in order to compute the three benefits accounted for within the procedure (savings in time, vehicle operating costs and accident costs) the program requires that the number of each category of vehicle utilising the link under examination throughout its economic life be determined using origin and destination data gathered from traffic surveys.

The inputs to the COBA analysis are hugely dependent on the output of the traffic forecasting and modelling process outlined in the previous chapter. It assumes a fixed demand matrix of trips based on knowledge of existing flows and available traffic forecasts where travel demands in terms of origin and destinations and modes and times of travel remain unchanged. This assumption has the advantage of being relatively simple to apply and has been used successfully for simple road networks. It has difficulty, however, in coping with complex networks in urban areas or in situations where congestion is likely to occur on links directly affecting the particular scheme being assessed. This has a direct effect on the traffic assignment stage of the traffic modelling process, which is of central importance to the proper working of the COBA program. In the case of complex urban networks, where urban schemes result in changes in travel behaviour that extend beyond simple reassignment of trips, more complex models such as UREKA have been developed to predict flows.

3.3.8   Advantages and disadvantages of cost-benefit analysis

The final output from a cost-benefit analysis, in the opinion of Kelso (1964), should be a cardinal number representing the dollar rate of the streams of net prime benefits of the proposal that he termed 'pure benefits'. Pure benefits measured the net benefits with the project in relation to net benefits without the project. Hill (1973) believed that this statement, one that explicitly sets out the basis for a traditional cost-benefit analysis, reveals some of the major deficiencies in the technique. Although there is some consideration of intangibles, they tend not to enter fully into the analysis. As a result, the effect of those investments that can be measured in monetary terms, whether derived directly or indirectly from the market, are implicitly treated as being more important, for the sole reason that they are measurable in this way, when in reality the intangible costs and benefits may have more significant consequences for the proposal. Furthermore, cost-benefit analysis is most suitable for ranking or evaluating different highway options, rather than for testing the absolute suitability of a project. This is, to an extent, because all valuations of costs and benefits are subject to error and uncertainty. Obtaining an absolute measure of suitability is an even greater limitation.

The advantages and disadvantages of cost-benefit analysis can be summarised as follows.

Advantages

- The use of the common unit of measurement, money, facilitates comparisons between alternative highway proposals and hence aids the decision-making process.
- Given that the focus of the method is on benefits and costs of the highway in question to the community as a whole, it offers a broader perspective than a narrow financial/investment appraisal concentrating only on the effects of the project on the project developers, be that the government or a group of investors funding a toll scheme.

Disadvantages

- The primary basis for constructing a highway project may be a societal or environmental rather than an economic one. If the decision is based solely on economic factors, however, an incorrect decision may result from the confusion of the original primary purpose of a proposed project with its secondary consequences, simply because the less important secondary consequences are measurable in money terms.
- The method is more suitable for comparing highway proposals designed to meet a given transport objective, rather than evaluating the absolute desirability of one project in isolation. This is partly because all estimates of costs

and benefits are subject to errors of forecasting. A decision-maker will thus feel more comfortable using it to rank a number of alternative highway design options, rather than to assess the absolute desirability of only one option relative to the existing 'do-nothing' situation, though this in some cases may be the only selection open to him/her.

- Although some limited recognition may be given to the importance of costs and benefits that cannot be measured in monetary terms, say, for example, the environmental consequences of the project in question, they tend to be neglected, or at best downgraded, within the main economic analysis. Those goods capable of measurement in monetary terms are usually attributed more implicit importance even though, in terms of the overall viability of the project, they may be less significant.

The first two disadvantages can be managed effectively by employing an experienced and competent decision expert to oversee use of the cost-benefit framework. Problems arising from the third point may require use of one of the other methodologies detailed later in the chapter. Some efforts have been made to provide monetary valuations for intangibles to enable their inclusion in cost-benefit. These techniques are in various stages of development.

3.4   Payback analysis

Payback analysis is an extremely simple procedure that is particularly useful in evaluating proposals such as privately funded highway projects where tolls will be imposed on users of the facility in order to recover construction costs. The method delivers an estimate of the length of time taken for the project to recoup its construction costs. It does not require information on an appropriate interest rate, but the lack of accuracy of the method requires that results from it should not be given the same weight as those from formal economic techniques outlined in this chapter, such as cost-benefit analysis. The method assumes that a given proposal will generate a stream of monies during its economic life, and at some point the total value of this stream will exactly equal its initial cost. The time taken for this equalisation to occur is called the payback period. It is more usefully applied to projects where the timescale for equalisation is relatively short. The method itself does not address the performance of the proposal after the payback period. Its analysis is thus not as complete as the more formal techniques, and therefore its results, when taken in isolation, may be misleading. It is therefore best utilised as a back-up technique, supplementing the information from one of the more comprehensive economic evaluation methods.

While the method has certain shortcomings, it is utilised frequently by engineering economists. Its strength lies in its simplicity and basic logic. It addresses a question that is very important to the developer of a tolled highway facility, as a relatively speedy payback will protect liquidity and release funds more quickly for investment in other ventures. This is particularly the case in times of recession when cash availability may be limited. Highway projects with a relatively short payback period can be attractive to a prospective developer. The short time frame is seen as lessening the risk associated with a venture, though road projects are seen as relatively low-risk enterprises.

The following formula enables the payback period to be derived:

$$\text{Payback period } (n_p) = (C_0 \div \text{NAS}) \tag{3.1}$$

where
C_0 = the initial construction cost of the highway project
NAS = net annual savings

Equation 3.1 assumes that a zero discount value is being used. This is not always the case. If it is assumed that the net cash flows will be identical from year to year, and that these cash flows will be discounted to present values using a value $i \neq 0$, then the uniform series present worth factor (P/A) can be utilised within the following equation:

$$0 = C_0 + \text{NAS}(P/A, i, n_p) \tag{3.2}$$

Equation 3.2 is solved to obtain the correct value of n_p.

The method is, however, widely used in its simplified form, with the discount rate, i, set equal to zero, even though its final value may lead to incorrect judgements being made. If the discount rate, i, is set equal to zero in Equation 3.2, the following relationship is obtained:

$$0 = C_0 + \sum_{t=1}^{t=n_p} \text{NAS}_t \tag{3.3}$$

Equation 3.3 reduces to $n_p = C_0/\text{NAS}$, exactly the same expression as given in Equation 3.1.

Example 3.1 – Comparison of toll-bridge projects based on payback analysis
A developer is faced with a choice between two development alternatives for a toll bridge project: one large-scale proposal with higher costs but enabling more traffic to access it, and the other less costly but with a smaller traffic capacity. Details of the costs and revenues associated with both are given in Table 3.5.

Calculate the payback period and check this result against the net present value for each.

Contd

Example 3.1 Contd

	Option A	Option B
Initial cost (£m)	27	50
Annual profit (£m)	5	10
Discount rate (%)	8	8
Life (years)	20	20

Table 3.5 Comparison of two options using payback analysis

Payback period

Option A $= C_0/\text{NAS} = 27/5 = 5.4$ years

Option B $= C_0/\text{NAS} = 50/8 = 6.25$ years

On the basis of simple payback, the cheaper option A is preferred to option B on the basis that the initial outlay is recouped in nearly one year less.

Present worth

The formula that converts an annualised figure into a present worth value, termed the series present worth factor (P/A), is expressed as:

$$P/A = \left((1+i)^n - 1\right)/\left(i(1+i)^n\right)$$

Assuming $i = 0.08$ and $n = 20$

$$P/A = \left((1.08)^{20} - 1\right)/\left(0.08(1.08)^{20}\right)$$
$$= 9818$$

therefore:

Present worth$_{\text{option A}} = -27 + 5 \times 9.818$
$$= +22.09$$

Present worth$_{\text{option B}} = -50 + 10 \times 9.818$
$$= +48.18$$

On the basis of its present worth valuation, option B is preferred, having a net present value over twice that of option A. Thus, while payback is a useful preliminary tool, primary methods of economic evaluation such as net present value or internal rate of return should be used for the more detailed analysis.

3.5 Environmental appraisal of highway schemes

While the cost-benefit framework for a highway project addresses the twin objectives of transport efficiency and safety, it makes no attempt to value its effects on the environment. Environmental evaluation therefore requires an alternative analytical structure. The structure developed within the last 30 years is termed environmental impact assessment (EIA).

The procedure has its origins in the US during the 1960s when environmental issues gained in importance. The legal necessity for public consultation during the planning stage of a highways project, allied to the preoccupation with environmental issues by environmental groups, resulted in the identified need for environmental assessment within the project planning process. The process was made statutory under the National Environmental Policy Act 1969 which requires the preparation of an environmental impact statement (EIS) for any environmentally significant project undertaken by the federal government. NEPA prescribes a format for the EIS, requiring the developer to assess:

- The probable environmental impact of the proposal
- Any unavoidable environmental impacts
- Alternative options to the proposal
- Short-run and long-run effects of the proposal and any relationship between the two
- Any irreversible commitment of resources necessitated by the proposal.

This list aids the identification and evaluation of all impacts relevant to the evaluation of the project concerned.

Interest in EIA spread to Europe during the 1970s in response to the perceived shortcomings within the then existing procedures for assessing the environmental consequences of large-scale development projects and for predicting the long-term direct and indirect environmental and social effects. The advantages of such a procedure was noted by the European Commission, and the contribution of EIA to proper environmental management was noted in the Second Action Programme on the Environment, published by them in 1977. A central objective of this programme was to put in place a mechanism for ensuring that the effects on the environment of development projects such as major highway schemes would be taken into account at the earliest possible stages within their planning process. A directive (85/337/EEC) (Council of the European Communities, 1985) giving full effect to these elements of European Union policy was agreed and passed in July 1985 with the requirement that it be transposed into the legislation of every member state within three years.

The directive helps ensure that adequate consideration is given to the environmental effects of a development project by providing a mechanism for ensuring that the environmental factors relevant to the project under examination are properly considered within a formal statement – the EIS – structured along broadly the same lines as the US model.

The directive also details the minimum information that must be contained within the EIS. These include:

(1) A physical description of the project
(2) A description of measures envisaged to reduce/remedy the significant
     adverse environmental effects of the project
(3) The data required to both identify and assess the main effects on the envi-
     ronment of the project in question.

Within the UK, since 1993 the *Design Manual for Roads and Bridges* (DoT,
1993) has provided the format within which the environmental assessment of
highway schemes has taken place. It identified 12 environmental impacts to be
assessed for any new/improved trunk road proposal. These, together with the
economic assessment, would form the decision-making framework used as the
basis both for choosing between competing options for a given highway route
corridor and for deciding in absolute terms whether the proposal in any form
should be proceeded with.

   The 12 environmental impacts forming the assessment framework are:

- *Air quality*   The main vehicle pollutants assessed are carbon monoxide
  (CO), oxides of nitrogen (NO_x) and hydrocarbons (HC), lead (Pb), carbon
  dioxide (CO_2) and particulates. Established models are used to predict future
  levels of these pollutants, and the values obtained are compared with current
  air quality levels.
- *Cultural heritage*   The demolition/disturbance of archaeological remains,
  ancient monuments and listed buildings and the impact of such actions on
  the heritage of the locality, are assessed under this heading.
- *Construction disturbance*   Though this impact is a temporary one, its effects
  can nonetheless be severe throughout the entire period of construction of
  the proposal. Nuisances such as dirt, dust, increased levels of noise and
  vibration created by the process of construction can be significant and may
  affect the viability of the project.
- *Ecology/nature conservation*   The highway being proposed may negatively
  affect certain wildlife species and their environment/habitats along the route
  corridor in question. Habitats may be lost, animals killed and flora/fauna
  may be adversely affected by vehicle emissions.
- *Landscape effects*   The local landscape may be fundamentally altered by the
  construction of the proposed highway if the alignment is not sufficiently
  integrated with the character of the local terrain.
- *Land use*   The effects of the route corridor on potential land use proposals
  in the area, together with the effects of the severance of farmlands and the
  general reduction, if any, in general property values in the vicinity of the
  proposed route, are assessed under this heading.
- *Traffic noise and vibration*   The number of vehicles using the road, the percentage of heavy vehicles, vehicle speed, the gradient of the road, the prevailing weather conditions and the proximity of the road to the dwellings
  where noise levels are being measured, all affect the level of noise nuisance

for those living near a road. Vehicle vibrations can also damage the fabric of buildings.

- *Pedestrian, cyclist and community effects*   The severance of communities and its effect on people in terms of increased journey time and the breaking of links between them and the services/facilities used daily by them, such as shops, schools and sporting facilities, are evaluated within this category of impact.
- *Vehicle travellers*   This assesses the proposal from the perspective of those using it, i.e. the drivers. The view from the road (scenery and landscape), the driver stress induced by factors such as the basic road layout and frequency of occurrence of intersections, are assessed within this category on the basis that they directly affect levels of driver frustration and annoyance leading to greater risk-taking by drivers.
- *Water quality and drainage*   This measures the effect that run-off from a road development may have on local water quality. Installations such as oil interceptors, sedimentation tanks and grit traps will, in most instances, minimise this effect, though special measures may be required in particular for water sources of high ecological value.
- *Geology and soils*   The process of road construction may destabilise the soil structure or expose hitherto protected rock formations. These potential impacts must be identified together with measures to minimise their effects.
- *Policies and plans*   This impact assesses the compatibility of the proposal with highway development plans at local, regional and national level.

Some of the above impacts can be estimated in quantitative terms, others only qualitatively. The exact method of assessment for each is detailed within the *Design Manual for Roads and Bridges*.

It is imperative that the environmental information is presented in as readily understandable a format as possible so that both members of the public and decision-makers at the highest political level can maximise their use of the information. One such format provided for in DMRB is the environmental impact table (EIT), a tabular presentation of data summarising the main impacts of a proposed highway scheme. At the early stages of the highway planning process, the EIT format can be used to consider alternative route corridors. As the process develops, specific routes will emerge and the level of environmental detail on each will increase. The 'do-nothing' scenario should also be considered as it defines the extent of the existing problem which has required the consideration of the development proposal in question. In most situations, the 'do-nothing' represents a deteriorating situation. If the baseline situation is to include localised highway improvements or certain traffic management measures, this option could more accurately be termed 'do-minimum'. DMRB advises that an EIT be constructed for all relevant appraisal groups. Three of these are:

- Local people and their communities
- Travellers (drivers and pedestrians)
- Cultural and national environment.

A listing of impacts relevant to each of these appraisal groups is given in Table 3.6. Table 3.7 gives an example of an EIT for group 1, local people and their communities.

Table 3.6 Appraisal groupings

Appraisal grouping	Impact
Local communities	Demolition of properties
	Noise
	Visual impact
	Severance
	Construction disruption
Travellers	Driver stress
	View from road
	Reduction in accidents
Cultural/national environment	Noise
	Severance
	Visual impact
	Landscape impacts

Table 3.7 Sample EIT for 'local people and their communities'

		Options	
Impact	Units	Preferred route	Do-minimum
Demolition of properties	Number	3	0
Noise	Number of properties experiencing an increase of more than		
	1 to 2	1	10
	3 to 4	3	2
	5 to 9	3	0
	10 to 14	5	0
	15 +	0	0
Visual impact	Number of properties subjected to visual impact		
	Substantial	1	0
	Moderate	2	4
	Slight	2	4
	No change	1	0
Severance	Number of properties		
	Obtaining relief to existing severance	2	0
	Having new severance imposed	3	0
Construction disruption	Number of properties within 100 m of site	3	0

In addition, a table listing the existing uses of land to be taken and a quantification of the specific areas required for the proposal should be included, together with a mitigation table listing the measures such as noise barriers, interceptors, balancing ponds and even local re-alignments proposed by the developer to minimise environmental impact.

3.6   The new approach to appraisal (NATA)

During the late 1990s, the UK government reviewed its road programme in England and identified those strategically important schemes capable of being started within the short to medium term and listed them as potential candidates for inclusion within a targeted programme.

Each of these schemes was subject to a new form of assessment that incorporated both the COBA-based economic appraisal and the EIT-based environmental assessment. This methodology, called the new approach to appraisal (NATA), includes a one-page summary of the impacts for each of the projects considered. Within the method, all significant impacts should be measured. Wherever possible, assessments should reflect the numbers affected in addition to the impact on each. It is desirable that all impacts be measurable in quantitative terms, though this may not always be feasible.

This appraisal summary table (AST) is designed for presentation to those decision-makers charged with determining whether approval for construction should be granted, and if so what level of priority should be assigned to it. It thus constitutes a key input into the process of scheme approval and prioritisation.

The AST summarises the assessment of the scheme in question against the following five objectives and their constituent impacts, seen by the government as being central to transport policy:

- *Environmental impact* Noise, air impacts, landscape, biodiversity, heritage and water
- *Safety*
- *Economy* Journey times, cost, journey time reliability, regeneration
- *Accessibility* Pedestrians, access to public transport, community severance
- *Integration.*

Environment

Noise

The impact of noise is quantified in terms of the number of properties whose noise levels in the year in question for the 'with proposal' option are greater or less than those in the base year. Given that only those properties subject to noise increases of greater than $3\,dB(A)$ are taken into account, the following quantities must be derived:

- The number of residential properties where noise levels within the assessment year for the 'with proposal' option are 3 dB(A) *lower* than for the 'do-minimum' option
- The number of residential properties where noise levels within the assessment year for the 'with proposal' option are 3 dB(A) *higher* than for the 'do-minimum' option.

Local air quality

Levels of both particulates PM_{10} (in micrograms per cubic metre) and nitrogen dioxide NO_2 (in parts per billion) are of particular concern. Firstly the roadside pollution levels for the year 2005 are identified for both the 'do-minimum' and 'with project' cases. Then the exposure to this change is assessed using the property count, with the diminishing contribution of vehicle emissions to pollution levels over distance taken into account using a banding of properties. The pollution increases of those dwellings situated nearer the roadside will receive a higher weighting than increases from properties further away under this system. Having separated out those parts of the route where air quality has improved and where it has worsened, for each affected section under examination a score for both PM_{10} and NO_2 are obtained:

Particulates score = (Difference in PM_{10} in 2005) × (weighted number of properties)

Nitrogen dioxide score = (Difference in NO_2 in 2005) × (weighted number of properties)

The final score is then obtained by aggregating the separate values across all affected sections. This computation is done separately for each pollutant.

In addition, the impacts of the proposals on global emissions are assessed using the net change in carbon dioxide levels as an overall indicator. To achieve this, the total forecast emissions after the proposal has been implemented are calculated and then deducted from the estimated values for the existing road network.

Landscape

NATA describes the character of the landscape and evaluates those features within it that are deemed important by the decision-maker. The result is a qualitative assessment, usually varying from large negative to slightly positive, with the intermediate points on the scale being moderately/slightly negative and neutral. In situations where the scheme is unacceptable in terms of visual intrusion, the assessment of 'very large negative' can be applied.

Biodiversity

The purpose of this criterion is to appraise the ecological impact of the road scheme on habitats, species or natural features. The appraisal summary table's standard seven-point scale (neutral, slight, moderate or large benefi-cial/adverse) is utilised. In situations where the scheme is unacceptable in terms of nature conservation, the assessment of 'very large negative' can be applied.

Heritage

This criterion assesses the impact of the proposal on the historic environment. It too is assessed on the AST's standard seven-point scale.

Water

In order to gauge the effect of the proposal on the water environment, a risk-based approach is adopted to assess its potentially negative impact on both water quality and land drainage. Both these are evaluated on a three-point scale of high/medium/low in an effort to gauge the overall sensitivity of the water environment. The potential of the proposal to cause harm is then determined using two indicators:

- Traffic flows – relating to water quality
- The surface area of the proposal (total land take) – relating to land drainage/flood defence.

Again, for this stage, the same three-point assessment scale is used (high/medium/low). In relation to water quality, traffic flows in excess of 30000 annual average daily traffic (AADT) are assessed as having a 'high' poten-tial to cause harm, with flows between 15000 and 30000 AADT assessed as 'medium' and those less than 15000 AADT assessed as 'low'. For land drainage/flood defence, areas in excess of 40 ha are assessed as having a 'high' potential to cause harm, with areas between 10 and 40 ha assessed as 'medium' and areas less than 10 ha assessed as 'low'. Based on the information from both stages, an assessment using only the neutral/negative points on the AST's assessment scale is used to indicate the proposal's overall performance on this criterion.

Safety

This criterion measures the extent to which the proposal improves the safety for travellers, indicating its effectiveness in terms of the monetary value, in present value terms, of the reduction in accidents brought about directly by the con-struction of the new/improved road. This requires accidents to be broken down into those causing death, those causing serious injury and those resulting in only slight injury. The results for this criterion can be obtained directly from COBA. The discount rate used is 6%, with all values given in 1994 prices, and it includes accidents likely to occur during both the construction and maintenance phases of the proposed road.

Economy   045211

The degree to which the proposal contributes both to economic efficiency and to sustainable economic growth in appropriate locations is assessed under this heading. A discount rate of 6% and a base year 1994 are again utilised for indicators assessed in monetary terms. Four indicators are used, as follows.

Journey times and vehicle operating costs
The effectiveness of the proposal on this criterion is measured in terms of the monetary value, in present value terms, of the reductions in both journey times and vehicle operating costs brought about directly by the construction of the new/improved road.

Costs
The present value of the costs of construction net of the cost of construction of the 'do-minimum' option.

Reliability
This assesses the impact of the proposal on the objective of improving the journey time reliability for road users. Reliability is reduced as flows reach capacity and stress levels increase. Stress can be defined as the ratio of the AADT to the congestion reference flow (CRF), expressed as a percentage. (CRF measures the performance of a link between junctions.) Reliability is not an issue for stress levels below 75%, with 125% as the upper limit. The assessment is based on the product of this percentage and the number of vehicles affected. The difference in stress for the old and new routes should be estimated. The final assessment is based on the product of flow and the difference in stress. Values in excess of +/− 3 million are classified as large (positive or negative), +/− 1 to 3 million classified as moderate, +/− 0.2 to 1 million classified as slight and values less than 0.2 million classified as neutral.

Regeneration
This evaluates whether the proposal is consistent with government regeneration objectives. The final assessment is a simple yes/no to this question, based on the extent to which the road is potentially beneficial for designated regeneration areas and on the existence of significant developments within or near regeneration areas likely to depend on the road's construction.

Accessibility

This criterion relates to the proposal's impact on the journeys made within the locality by modes of transport other than the private car, assessing whether the proposed project will make it easier or more difficult for people to journey to work by public transport, on foot, by bicycle or other means.

Pedestrians, cyclists and equestrians
This subcriterion relates to the proposal's impact on the journeys made within the locality on foot (pedestrian), by bicycle (cyclist) or by horse (equestrian). The assessment should be based on the year of opening, taking typical daily conditions.

First, a quantitative assessment of the change in accessibility for each group is estimated by multiplying together the numbers in the grouping affected, the change in journey time (in minutes) and the change in amenity (+1, −1 or 0 depending on whether accessibility has been improved, worsened or has remain unchanged). The three valuations are then added together to give an overall score. The final assessment is given using the standard AST seven-point scale:

- *Beneficial* – journey times reduced
- *Adverse* – journey times increased
- *Slight* – fewer than 200 travellers affected, journey times are changed by less than 1 minute and there is no change in amenity
- *Large* – typically, more than 1000 travellers are affected, journey times are changed significantly (by more than 1 minute) and there are changes in amenity
- The assessment in all intermediate cases will be *Moderate*.

Access to public transport
The extent to which access to public transport by non-motorised modes is affected by the proposal is assessed within this heading. Broadly the same framework as above is used, with the score on the seven-point AST scale based on the number of public transport users affected, the changes in access time to the service and the degree to which the quality of the service would be improved (+1), made worse (−1) or unaffected (0) as a result of the proposal under examination.

Community severance
The severance effect on those travellers using non-motorised modes is assessed on the standard AST seven-point scale:

- *Beneficial* – relief from severance
- *Adverse* – new severance
- *Neutral* – new severance is balanced by relief of severance (the net effect is approximately zero)
- *Slight* – low level severance with very few people affected (less than 200)
- *Large* – severe level severance with many people affected (more than 1000)
- The assessment in all intermediate cases will be *Moderate*.

Integration

This criterion assesses in broad terms the compatibility of the proposal with land use and transportation plans and policies at local, regional and national level. A three-point textual scale (neutral-beneficial-adverse) is used:

- *Beneficial* – more policies are facilitated than hindered by the construction of the proposal
- *Adverse* – more policies are hindered than facilitated by the construction of the proposal
- *Neutral* – the net effect on policies is zero.

This assessment is intended to be broad-brush in approach, with marginal changes ignored.

The AST framework is summarised in Table 3.8.

Table 3.8 Framework for appraisal summary table

Criterion	Subcriterion	Quantitative measure	Assessment
Environment	Noise	Number of properties experiencing: • An increase in noise levels • A decrease in noise levels	Net number of properties who win with scheme
	Local air quality	Number of properties experiencing: • Improved air quality • Worse air quality	PM_{10} score NO_2 score
	Landscape	—	AST 7-point scale
	Biodiversity	—	AST 7-point scale
	Heritage	—	AST 7-point scale
	Water	—	AST 7-point scale
Safety		Number of deaths, serious injuries and slight injuries	Present value of the benefits (PVB) due to accident reductions (£m)
Economy	Journey times and VOCs	—	Present value of the benefits (PVB) due to journey time and vehicle operating cost savings (£m)
	Cost	—	Present value of the costs (PVC) of construction (£m)
	Reliability	% stress before and after project implementation	Four-point scale Large/Moderate/Slight/Neutral
	Regeneration	Does proposal serve a regeneration priority area? Does regeneration depend on the construction of the proposal?	Yes/no
Accessibility	Pedestrians, etc.	—	AST 7-point scale
	Public transport	—	AST 7-point scale
	Severance	—	AST 7-point scale
Integration		Consistent with implementation of local/regional/national development plans	Beneficial/Neutral/Adverse

3.7   Summary

This chapter summarises the main types of methodologies for assessing the desirability both in economic and in environmental/social terms of constructing a highway proposal. While the economic techniques may have been the first to gain widespread acceptance, there is now a broad awareness, both within the UK and in Europe as a whole, as well as the US, that highway appraisal must be as inclusive a process as possible. Such concerns were the catalyst for the introduction of the environmental impact assessment process. This inclusiveness requires that the deliberations of as many as possible of the groupings affected by the proposed scheme should be sought, and that the scheme's viability should be judged on as broad a range of objectives/criteria as possible.

3.8   References

Adler, H.A. (1987) *Economic Appraisal of Transport Projects: A Manual with Case Studies.* EDI Series in Economic Development, Johns Hopkins University Press, London (Published for the World Bank).

Council of the European Communities (1985) On the assessment of the effects of certain public and private projects on the environment. *Official Journal L175*, 28.5.85, 40–48 (85/337/EEC).

DoT (1982) Department of Transport *COBA: A method of economic appraisal of highway schemes.* The Stationery Office, London.

DoT (1993) Department of Transport *Design Manual for Roads and Bridges, Vol. 11: Environmental Impact Assessment.* The Stationery Office, London.

Hill, M. (1973) *Planning for Multiple Objectives: An Approach to the Evaluation of Transportation Plans.* Technion, Philadelphia, USA.

Kelso, M.M. (1964) Economic analysis in the allocation of the federal budget to resource development. In *Economics and Public Policy in Water Resource Development* (eds S.C. Smith & E.N. Castle) pp. 56–82. Iowa State University Press, USA.

Martin, J. C. (1993) *The Successful Engineer: Personal and Professional Skills – A Sourcebook.* McGraw-Hill International Editions, New York, USA.

Riggs, J.L., Bedworth, D.D. & Randhawa, S.U. (1997) *Engineering Economics.* McGraw-Hill International Editions, New York, USA.

Department of the Environment, Transport and the Regions DETR (1998) *A Guidance on the New Approach to Appraisal (NATA)*, September. The Stationery Office, London.

Zey, M. (1992) Criticisms of rational choice models. In *Decision Making: Alternatives to Rational Choice Models.* Sage, Newbury Park, California, USA.

Chapter 4

Basic Elements of Highway Traffic Analysis

4.1 Introduction

The functional effectiveness of a highway is measured in terms of its ability to assist and accommodate the flow of vehicles with both safety and efficiency. In order to measure its level of effectiveness, certain parameters associated with the highway must be measured and analysed. These properties include:

- The quantity of traffic
- The type of vehicles within the traffic stream
- The distribution of flow over a period of time (usually 24 hours)
- The average speed of the traffic stream
- The density of the traffic flow.

Analysis of these parameters will directly influence the scale and layout of the proposed highway, together with the type and quantity of materials used in its construction. This process of examination is termed traffic analysis and the sections below deal with relationships between the parameters which lie at its basis.

4.2 Speed, flow and density of a stream of traffic

The traffic flow, q, a measure of the volume of traffic on a highway, is defined as the number of vehicles, n, passing some given point on the highway in a given time interval, t, i.e.:

$$q = \frac{n}{t} \tag{4.1}$$

In general terms, q is expressed in vehicles per unit time.

The number of vehicles on a given section of highway can also be computed in terms of the density or concentration of traffic as follows:

$$k = \frac{n}{l} \tag{4.2}$$

where the traffic density, k, is a measure of the number of vehicles, n, occupying a length of roadway, l.

For a given section of road containing k vehicles per unit length l, the average speed of the k vehicles is termed the space mean speed u (the average speed for all vehicles in a given space at a given discrete point in time).

Therefore:

$$u = \frac{(1/n)\sum_{i=1}^{n} l_i}{t} \qquad (4.3)$$

where l_i is the length of road used for measuring the speed of the ith vehicle.

It can be seen that if the expression for q is divided by the expression for k, the expression for u is obtained:

$$q \div k = \left[\frac{n}{t}\right] \div \left[\frac{n}{l}\right] = \left[\frac{n}{t}\right] \times \left[\frac{l}{n}\right] = \frac{l}{t} = u \qquad (4.4)$$

Thus, the three parameters u, k and q are directly related under stable traffic conditions:

$$q = uk \qquad (4.5)$$

This constitutes the basic relationship between traffic flow, space mean speed and density.

4.2.1   Speed-density relationship

In a situation where only one car is travelling along a stretch of highway, densities (in vehicles per kilometre) will by definition be near to zero and the speed at which the car can be driven is determined solely by the geometric design and layout of the road; such a speed is termed free-flow speed as it is in no way hindered by the presence of other vehicles on the highway. As more vehicles use the section of highway, the density of the flow will increase and their speed will decrease from their maximum free-flow value (u_f) as they are increasingly more inhibited by the driving manoeuvres of others. If traffic volumes continue to increase, a point is reached where traffic will be brought to a stop, thus speeds will equal zero ($u = 0$), with the density at its maximum point as cars are jammed bumper to bumper (termed jam density, k_j).

Thus, the limiting values of the relationship between speed and density are as follows:

When $k = 0$, $u = u_f$
When $u = 0$, $k = k_j$

Various attempts have been made to describe the relationship between speed

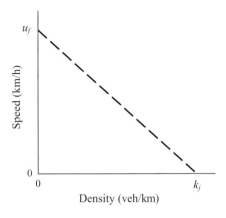

Figure 4.1
Illustration of speed-density relationship.

and density between these two limiting points. Greenshields (1934) proposed the simplest representation between the two variables, assuming a linear relationship between the two (see Fig. 4.1).

In mathematical terms, this linear relationship gives rise to the following equation:

$$u = u_f \left(1 - \frac{k}{k_j} \right) \tag{4.6}$$

This assumption of linearity allows a direct mathematical linkage to be formed between the speed, flow and density of a stream of traffic.

This linear relationship between speed and density, put forward by Greenshields (1934), leads to a set of mathematical relationships between speed, flow and density as outlined in the next section. The general form of Greenshields' speed-density relationship can be expressed as:

$$u = c_1 + c_2 k \tag{4.7}$$

where c_1 and c_2 are constants.

However, certain researchers (Pipes, 1967; Greenberg, 1959) have observed non-linear behaviour at each extreme of the speed-density relationship, i.e. near the free-flow and jam density conditions. Underwood (1961) proposed an exponential relationship of the following form:

$$u = c_1 \exp(-c_2 k) \tag{4.8}$$

Using this expression, the boundary conditions are:

- When density equals zero, the free flow speed equals c_1
- When speed equals zero, jam density equals infinity.

The simple linear relationship between speed and density will be assumed in all the analyses below.

4.2.2   Flow-density relationship

Combining Equations 4.5 and 4.6, the following direct relationship between flow and density is derived:

$$q = uk = u_f\left(1 - \frac{k}{k_j}\right) \times k, \text{ therefore}$$

$$q = u_f\left(k - \frac{k^2}{k_j}\right) \tag{4.9}$$

This is a parabolic relationship and is illustrated below in Fig. 4.2.

In order to establish the density at which maximum flow occurs, Equation 4.9 is differentiated and set equal to zero as follows:

$$\frac{dq}{dt} = u_f\left(1 - \frac{2k}{k_j}\right) = 0$$

since $u_f \neq 0$, the term within the brackets must equal zero, therefore:

$$1 - \frac{2k_m}{k_j} = 0, \text{ thus}$$

$$k_m = \frac{k_j}{2} \tag{4.10}$$

k_m, the density at maximum flow, is thus equal to half the jam density, k_j. Its location is shown in Fig. 4.2.

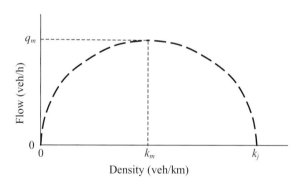

Figure 4.2
Illustration of flow-density relationship.

4.2.3   Speed-flow relationship

In order to derive this relationship, Equation 4.6 is rearranged as:

$$k = k_j\left(1 - \frac{u}{u_f}\right) \tag{4.11}$$

By combining this formula with Equation 4.5, the following relationship is derived:

$$q = k_j \left(u - \frac{u^2}{u_f} \right)$$ (4.12)

This relationship is again parabolic in nature. It is illustrated in Fig. 4.3.

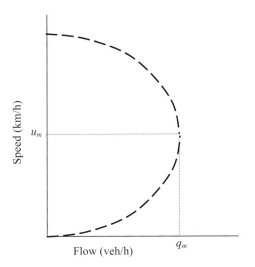

Figure 4.3
Illustration of speed-flow relationship.

In order to find the speed at maximum flow, Equation 4.12 is differentiated and put equal to zero:

$$\frac{dq}{dt} = k_j \left(1 - \frac{2u}{u_f} \right) = 0$$

since $k_j \neq 0$, the term within the brackets must equal zero, therefore:

$$1 - \frac{2u_m}{u_f} = 0, \text{ thus}$$

$$u_m = \frac{u_f}{2}$$ (4.13)

u_m, the speed at maximum flow, is thus equal to half the free-flow speed, u_f. Its location is shown in Fig. 4.3.

Combining Equations 4.10 and 4.13, the following expression for maximum flow is derived:

$$q_m = u_m \times k_m = \frac{u_f}{2} \times \frac{k_j}{2}$$

therefore

$$q_m = \frac{u_f k_j}{4}$$ (4.14)

Example 4.1

Two platoons of cars are timed over a distance of 0.5 km. Their flows are recorded. The first group is timed at 40 seconds, with the flow at 1350 vehicles per hour. The second group take 45 seconds, with a flow of 1800 vehicles per hour.

Determine the maximum flow of the traffic stream.

Solution

Group 1 has an average speed of 45 km/h
Group 2 has an average speed of 40 km/h
Group 1 k value = 1350/45 = 30 v/km
Group 2 k value = 1800/40 = 45 v/km

To get the consequent relationship between speed and density based on the above two results, use co-ordinate geometry:

$$y - y_1 = m(x - x_1)$$

where

$$m = \frac{y_1 - y_2}{x_1 - x_2}$$

$$y = \text{speed}$$
$$x = \text{density}$$

The slope, *m*, of the line joining the above two results = −5/15 = −1/3

$$y - 45 = -1/3(x - 30)$$
$$y + x/3 = 45 + 10$$
$$y + x/3 = 55$$

Examining the boundary conditions:

Free flow speed = 55 km/h
Jam density = 165 v/km
Max flow = 55 * 165/4 = 2269 v/h

4.3   Determining the capacity of a highway

There are two differing approaches to determining the capacity of a highway. The first, which can be termed the 'level of service' approach, involves establishing, from the perspective of the road user, the quality of service delivered by a highway at a given rate of vehicular flow per lane of traffic. The methodology is predominant in the US and other countries. The second approach, used in Britain, puts forward practical capacities for roads of various sizes and width carrying different types of traffic. Within this method, economic assessments are used to indicate the lower border of a flow range, the level at which a given road width is likely to be preferable to a narrower one. An upper limit is also arrived at using both economic and operational assessments. Together these boundaries indicate the maximum flow that can be accommodated by a given carriageway width under given traffic conditions.

4.4   The 'level of service' approach

4.4.1   Introduction

'Level of service' describes in a qualitative way the operational conditions for traffic from the viewpoint of the road user. It gauges the level of congestion on a highway in terms of variables such as travel time and traffic speed.

The Highway Capacity Manual in the US (TRB, 1985) lists six levels of service ranging from A (best) to F (worst). There are each defined briefly as follows:

Service A: This represents free-flow conditions where traffic flow is virtually zero. Only the geometric design features of the highway, therefore, limit the speed of the car. Comfort and convenience levels for road users are very high as vehicles have almost complete freedom to manoeuvre.

Service B: Represents reasonable free-flow conditions. Comfort and convenience levels for road users are still relatively high as vehicles have only slightly reduced freedom to manoeuvre. Minor accidents are accommodated with ease although local deterioration in traffic flow conditions would be more discernible than in service A.

Service C: Delivers stable flow conditions. Flows are at a level where small increases will cause a considerable reduction in the performance or 'service' of the highway. There are marked restrictions in the ability to manoeuvre and care is required when changing lane. While minor incidents can still be absorbed, major incidents will result in the formation of queues. The speed chosen by the driver is substantially affected by that of the other vehicles. Driver comfort and convenience have decreased perceptibly at this level.

Service D: The highway is operating at high-density levels but stable flow still prevails. Small increases in flow levels will result in significant operational difficulties on the highway. There are severe restrictions on a driver's ability to manoeuvre, with poor levels of comfort and convenience.

Service E: Represents the level at which the capacity of the highway has been reached. Traffic flow conditions are best described as unstable with any traffic incident causing extensive queuing and even breakdown. Levels of comfort and convenience are very poor and all speeds are low if relatively uniform.

Service F: Describes a state of breakdown or forced flow with flows exceeding capacity. The operating conditions are highly unstable with constant queuing and traffic moving on a 'stop-go' basis.

These operating conditions can be expressed graphically with reference to the basic speed-flow relationship, as illustrated in Fig. 4.3. At the level of service A, speed is near its maximum value, restricted only by the geometry of the road, and flows are low relative to the capacity of the highway, given the small number of vehicles present. At the level of service D, flows are maximised, with speed at approximately 50% of its maximum value. Level of service F denotes the 'breakdown' condition at which both speeds and flow levels tend towards zero.

These conditions and their associated relative speeds and flows are illustrated in Fig. 4.4.

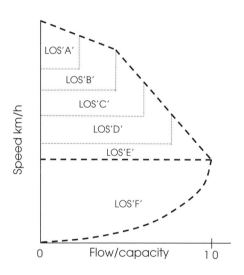

Figure 4.4 Linkage between level of service (LOS), speed and flow/capacity.

4.4.2   *Some definitions*

In order to determine a road's level of service, a comprehension of the relationship between hourly volume, peak hour factor and service flow is vital:

Hourly volume (V) The highest hourly volume within a 24-hour period

Peak-hour factor (PHF) The ratio of the hourly volume to the peak 15 minute flow (V_{15}) enlarged to an hourly value

$$\text{PHF} = V \div V_{15} \times 4 \tag{4.15}$$

Service flow (SF) The peak 15 minute flow (V_{15}) enlarged to an hourly value

$$\text{SF} = V_{15} \times 4 \tag{4.16}$$

4.4.3   Maximum service flow rates for multi-lane highways

The Highway Capacity Manual generates maximum flow values obtainable on a multi-lane highway given a certain speed limit and prevailing level of service. The values assume that ideal conditions exist, i.e. all carriageways are a standard width (3.65 m), there are no obstructions within 3.65 m of their edge, there are no heavy goods vehicles, buses or recreational vehicles on the road, the driver population consists of regular weekday drivers and the road is divided by a physical barrier and rural-based.

Given the existence of ideal conditions, the maximum service flow, $SF_{Max(i)}$, can be defined as:

$$SF_{Max(i)} = C_j \times \left(\frac{v}{c}\right)_i \times N \qquad (4.17)$$

N is the number of lanes in each direction, and C_j is the capacity of a standard highway lane for a given design speed j. Its values are shown in Table 4.1:

	Design speed (km/h)		
	70	60	50
C_j (v/h)	2000	2000	1900

Table 4.1 Values of C_j for different design speeds (Source: *Highway Capacity Manual* (TRB, 1985))

The maximum ratios of flow to capacity for each level of service and design speed limit are given in Table 4.2.

Level of service	$v/c(C_{70})$	$v/c(C_{60})$	$v/c(C_{50})$
A	0.36	0.33	—
B	0.54	0.50	0.45
C	0.71	0.65	0.60
D	0.87	0.80	0.76
E	1.0	1.00	1.00
F	Variable	Variable	Variable

Table 4.2 Ratios of flow to capacity for different levels of service and design speeds (Source: *Highway Capacity Manual* (TRB, 1985))

Example 4.2
A rural divided 4-lane highway has a peak hour volume (V) in one direction of 1850 vehicles per hour. Ideal conditions apply, therefore there are no heavy goods vehicles, buses or recreational vehicles in the traffic. The peak hour factor is 0.8. The design speed limit is 70 mph.

Determine the level of service being provided by the highway.

Contd

> **Example 4.2 Contd**
>
> *Solution*
>
> The service flow can be calculated knowing the hourly volume during the peak hour and the peak hour factor:
>
> $$\text{SF} = V \div \text{PHF}$$
> $$= 1850 \div 0.8 = 2312.5 \text{ vehicles per hour}$$
> $$C_{70} = 2000 \text{ passenger cars per hour per lane}$$
> $$N \text{ (the number of lanes in each direction)} = 2$$
>
> Since
>
> $$\text{SF}_{Max(i)} = C_j \times \left(\frac{v}{c}\right)_i$$
>
> Therefore
>
> $$\left(\frac{v}{c}\right)_i = \text{SF} \div C_j = 2312.5 \div (2000 \times 2) = 0.58$$
>
> Under the prevailing ideal conditions, therefore, with reference to Table 4.2, the ratio of flow to capacity is greater than 0.54 but less than 0.71. The highway thus provides level of service C.

For non-ideal conditions, Equation 4.17 becomes the following:

$$\text{SF}_{(i)} = C_j \times \left(\frac{v}{c}\right)_i \times N \times f_w \times f_{hv} \times f_p \times f_E \tag{4.18}$$

When lane widths are narrower than 3.65 m and/or barriers, lighting posts or any such obstructions are closer than 1.83 m from the edge of the travelled pavement (either at the kerb or median), an adjustment factor f_w must be introduced.

If the lane width is reduced to 2.74 m (9 ft) and there are obstructions at both edges bounding it, the capacity will be reduced by 34%, or just over one-third. Table 4.3 gives the adjustment factors for a 4-lane divided multi-lane highway.

Figures can also be obtained from the Highway Capacity Manual for 2-lane undivided, 4-lane undivided and 6-lane divided and undivided highways.

Heavy vehicles such as trucks, buses and recreational vehicles have a negative effect on the capacity of a highway due to their physical size together with their relatively slow acceleration and braking. The resulting reduction in capacity, termed the f_{HV} correction, is estimated on the basis of the amount of road space taken up by each of these vehicle types relative to that taken up by a private car combined with the percentage of such vehicles in the traffic stream in question.

Table 4.3 Correction factors for non-ideal lane widths and clearances from obstructions (multi-lane highways) (Source: *Highway Capacity Manual* (TRB, 1985))

Distance of obstruction from travelled edge (m)	Adjustment factor, f_w							
	Obstruction on one side of roadway				Obstruction on both sides of roadway			
	Lane width (m)				Lane width (m)			
	3.65 m	3.36 m	3.05 m	2.75 m	3.65 m	3.36 m	3.05 m	2.75 m
1.83 or greater	1.00	0.97	0.91	0.81	1.00	0.97	0.91	0.81
1.22	0.99	0.96	0.90	0.80	0.98	0.95	0.89	0.79
0.61	0.97	0.94	0.88	0.79	0.94	0.91	0.86	0.76
0	0.90	0.87	0.82	0.73	0.81	0.79	0.74	0.66

The passenger car equivalent (pce), or the number of equivalent private cars that would occupy the same quantity of road space, for each of the above types of heavy vehicle is primarily dependent on the terrain of the highway under examination, with steep gradients magnifying the performance constraints of the heavy vehicles.

The pce's for trucks (E_T), buses (E_B), and recreational vehicles (E_R), are defined for three different classes of terrain:

Level terrain: This is categorised as gradients or horizontal/vertical alignments that allow heavy vehicles to maintain the same speeds as private cars. Upward and downward gradients of not more than 1–2 % are normally consistent with this classification.

Rolling terrain: Those gradients or horizontal alignments that result in the speed of the heavy vehicle in question being lowered to a value substantially below those of the private car on the same stretch of roadway. The heavy vehicle is not operating at its maximum speed for a substantial distance.

Mountainous terrain: Those gradients or horizontal alignments that result in the heavy vehicle operating at its maximum speed for a substantial distance.

Values given by the Transportation Research Board are noted in Table 4.4.

Correction factor	Type of terrain		
	Level	Rolling	Mountainous
E_T for trucks	1.7	4.0	8.0
E_B for buses	1.5	3.0	5.0
E_R for RVs	1.6	3.0	4.0

Table 4.4 Passenger car equivalents for different classes of heavy vehicles (Source: *Highway Capacity Manual* (TRB, 1985))

Where the road gradient is greater than 3% over a distance of $\frac{1}{2}$ mile or less than 3% but over 2% over a distance greater than 1 mile, these values are no longer valid and more detailed tables as presented in the *Highway Capacity Manual* (TRB, 1985) must be utilised.

Having obtained the necessary pce valuations, the overall correction factor can be estimated once the percentages of the three vehicle types present along the section of road in question has been arrived at:

P_T – Percentage of trucks in traffic stream
P_B – Percentage of buses in traffic stream
P_R – Percentage of recreational vehicles in traffic stream.

Given these values, the correction factor, f_{HV}, can be derived as follows:

$$f_{HV} = \frac{1}{1 + \{P_T(E_T - 1) + P_B(E_B - 1) + P_R(E_R - 1)\}} \tag{4.19}$$

If the driver population is deemed not to be ideal, i.e. not composed entirely of regular weekday commuters, then a reduction factor can be utilised, reducing the capacity of the highway by anything between 10% and 25%. There are no quantitatively derived guidelines that can assist in making this assessment. Professional judgement must be the basis for the valuation used. The value range is illustrated in Table 4.5.

With regard to the type of highway, the ideal situation is represented by a divided highway in a rural setting. If, however, the highway is undivided and/or the setting is urban based, a correction factor must be used to take account of the resulting reduction in capacity. This correction factor, f_E, reflects the reduction in capacity resulting from the absence of a physical barrier along the centre-line of the road, with consequent interference from oncoming traffic together with the greater likelihood of interruptions in the traffic stream in an urban or suburban environment.

Values of f_E given in the *Highway Capacity Manual* are shown in Table 4.6.

Table 4.5 Correction factors for driver population types (Source: *Highway Capacity Manual* (TRB, 1985))

Driver classification	Correction factor, f_p
Regular weekday commuters	1.0
Other classes of drivers	0.9–0.75

Table 4.6 Correction factors for highway environment (Source: *Highway Capacity Manual* (TRB, 1985))

Highway classification	Divided	Undivided
Rural	1.0	0.95
Urban/suburban	0.9	0.80

Example 4.3

A suburban undivided 4-lane highway on rolling terrain has a peak hour volume (V) in one direction of 1500 vehicles per hour, with a peak hour factor estimated at 0.85. All lanes are 3.05 m (10 ft) wide. There are no obstructions within 1.83 m (6 ft) of the kerb.

The percentages for the various heavy vehicle types are:

$P_T - 12\%$
$P_B - 6\%$
$P_R - 2\%$

Determine the level of service of this section of highway.

Solution

The service flow is again calculated knowing the hourly volume during the peak hour and the peak hour factor:

$$SF = V \div PHF$$
$$= 1500 \div 0.85 = 1764.71 \text{ vehicles per hour}$$

C_{60} = 2000 passenger cars per hour per lane N (the number of lanes in each direction) = 2

f_w = 0.91 (3.05 m wide lanes, no roadside obstructions)

$$f_{HV} = \frac{1}{1 + \{P_T(E_T - 1) + P_B(E_B - 1) + P_R(E_R - 1)\}}$$

$$\frac{1}{1 + \{0.12(3) + 0.06(2) + 0.02(2)\}} = 0.66$$

f_P = 1.0 (ideal driver population)
f_E = 0.80 (suburban undivided)

Since

$$SF_{(i)} = C_j \times \left(\frac{v}{c}\right)_i \times N \times f_w \times f_{hv} \times f_p \times f_E$$

Therefore

$$\left(\frac{v}{c}\right)_i = SF_i \div (C_j \times N \times f_w \times f_{HV} \times f_p \times f_E)$$

$$\left(\frac{v}{c}\right)_i = 1764.71 \div (2000 \times 2 \times 0.91 \times 0.66 \times 1.0 \times 0.8) = 0.92 \qquad (4.20)$$

Using the data from Table 4.2, the highway operates at level of service E.

4.4.4   Maximum service flow rates for 2-lane highways

Where one lane is available for traffic in each direction, a classification of 2-lane highway applies. In such a situation, if a driver wishes to overtake a slower moving vehicle, the opposing lane must be utilised. This manoeuvre is therefore subject to geometric constraints, most noticeably passing sight distances but also the terrain of the stretch of road in question.

The capacity of such highways is expressed as a two-directional value rather than the one-directional value used in the previous section for multi-lane highways. Under ideal conditions, the capacity of a 2-lane highway is set at 2800 passenger car units per hour. If one adjusts this value by a ratio of flow to capacity consistent with the desired level of service, the following formula for service flow is obtained:

$$SF_i \div 2800 \times \left(\frac{v}{c}\right)_i \tag{4.21}$$

Ideal conditions assume the following:

- Passing is permissible throughout 100% of the section of highway in question
- Lane widths are at least 12 ft (3.65 m)
- Clearance on hard shoulders of at least 6 ft (1.83 m)
- Design speed of a least 60 miles per hour (96 km/h)
- The traffic stream entirely composed of private cars
- The flow in both directions exactly evenly balanced (50/50 split)
- Level terrain
- No obstructions to flow caused by vehicle turning movements, traffic signalisation, etc.

If ideal conditions obtain, the service flow is obtained using the ratios of flow to capacity associated with the required level of service, as given in Table 4.7.

When conditions are non-ideal, the capacity of the highway reduces from 2800 pcu/hour based on the following equation:

$$SF_i \div 2800 \times \left(\frac{v}{c}\right)_i \times f_d \times f_w \times f_{HV} \tag{4.22}$$

Level of service	Average speed	v/c ratio
A	≥58	0.15
B	≥55	0.27
C	≥52	0.43
D	≥50	0.64
E	≥45	1.0
F	<45	—

Table 4.7 Level of service values for 2-lane highways under ideal conditions (average travel speeds assume a design speed of 60 mph applies) (Source: *Highway Capacity Manual* (TRB, 1985))

All these correction factors are as defined in the previous sections except for f_d which takes account of any unequal distribution of the traffic between the two directions of flow.

The capacity of the flow in the peak direction decreases as the distribution becomes more unequal. The value for different flow splits between 50/50 and 100/0 are given in Table 4.8.

Table 4.8 Correction factor for different directional splits on a 2-lane highway (Source: *Highway Capacity Manual* (TRB, 1985))

Distributional split	50/50	60/40	70/30	80/20	90/10	100/0
Correction factor, f_d	1.0	0.94	0.89	0.83	0.75	0.71

Therefore, if all the flow is in the peak direction, the capacity is effectively 71% of 2800 pcu/h, i.e. just under 2000 pcu/h.

The correction factor for lane widths/clearance on hard shoulders also varies with the incident level of service. Typical values are given in Table 4.9.

Clearance on hard shoulder (metres)	3.65 m lanes (12 ft)		2.75 m lanes (9 ft)	
	LOS A-D	LOS E	LOS A-D	LOS E
1.83 or greater	1.00	1.00	0.70	0.76
1.22	0.92	0.97	0.65	0.74
0.61	0.81	0.93	0.57	0.70
Zero	0.70	0.88	0.49	0.66

Table 4.9 Correction factors for non-ideal lane widths and clearances from obstructions (2-lane highways) (Source: *Highway Capacity Manual* (TRB, 1985))

Intermediate values for 11 ft and 10 ft lane widths are available

The correction factor for heavy vehicles also depends on the incident level of service. Typical values are given in Table 4.10.

Vehicle type	Level of service	Type of terrain		
		Level	Rolling	Mountainous
Trucks, E_T	A	2.0	4.0	7.0
	B or C	2.2	5.0	10.0
	D or E	2.0	5.0	12.0
Buses, E_B	A	1.8	3.0	5.7
	B or C	2.0	3.4	6.0
	D or E	1.6	2.9	6.5
RV's, E_R	A	2.2	3.2	5.0
	B or C	2.5	3.9	5.2
	D or E	1.6	3.3	5.2

Table 4.10 Passenger car equivalents for different classes of heavy vehicles (2-lane highways) (Source: Werner and Morrall (1976))

The level of service values in Table 4.7 assumes that passing is available over all sections of the road and that the terrain is level. If this is not the case and passing is prohibited on portions of the highway and/or the terrain is rolling or mountainous, the ratios of volume to capacity for the different levels of service must be adjusted accordingly. Ratios for the different terrain types and for zero, 40% and 100% no-passing zones are given in Table 4.11.

Table 4.11 Level of service values for 2-lane highways for different percentage no-passing zones (Source: *Highway Capacity Manual* (TRB, 1985))

		v/c ratio										
	Level terrain				Rolling terrain				Mountainous terrain			
		% no-passing				% no-passing				% no-passing		
LOS	Avg. speed*	0	40	100	Avg. speed*	0	40	100	Avg. speed*	0	40	100
A	≥58	0.15	0.09	0.04	≥57	0.15	0.07	0.03	≥56	0.14	0.07	0.01
B	≥55	0.27	0.21	0.16	≥54	0.26	0.19	0.13	≥54	0.25	0.16	0.10
C	≥52	0.43	0.36	0.32	≥51	0.42	0.35	0.28	≥49	0.39	0.28	0.16
D	≥50	0.64	0.60	0.57	≥49	0.62	0.52	0.43	≥45	0.58	0.45	0.33
E	≥45	1.00	1.00	1.00	≥40	0.97	0.92	0.90	≥35	0.91	0.84	0.78
F	<45	—	—	—	<40	—	—	—	<35	—	—	—

*Average speeds in miles per hour, intermediate ratios for 20%, 60% and 80% no-passing zones are available.

Example 4.4

A 2-lane highway has lane widths of 9 ft (2.75 m), with 6 ft (1.83 m) clear hard shoulders. There are no-passing zones along 40% of its length. The directional split is 70/30 in favour of the peak direction.

The percentages for the various heavy vehicle types are:

P_T – 10%
P_B – 4%
P_R – 2%

The terrain is rolling.

Calculate the service flow of the highway when running at full capacity.

Contd

Example 4.4 Contd

Solution

Since the road is operating at capacity, the level of service is assumed to be E.

Therefore:

f_w = 0.76 (2.75 m wide lanes, no roadside obstructions)
f_d = 0.89 (70/30 distributional split)
Rolling terrain, level of service E, therefore E_T = 5.0, E_B = 2.9 and E_R = 3.3

Therefore:

$$f_{HV} = \frac{1}{1+\{0.10(4)+0.04(1.9)+0.02(2.3)\}} = 0.657$$

$$SF_E = 2800 \times \left(\frac{v}{c}\right)_E \times f_w \times f_{HV} \times f_d$$

$$SF_E = 2800 \times 0.92 \times 0.76 \times 0.657 \times 0.89 = 1145 \text{ v/h}$$

Example 4.5
Determine the level of service provided by a 2-lane highway with a peak-hour volume (V) of 1200 and a peak-hour factor of 0.8. No passing is permitted on the highway. The directional split is 60/40 in favour of the peak direction. Both lanes are 12 ft (3.65 m) wide. There is a 1.22 m (4 ft) clearance on both hard shoulders.
    The percentages for the various heavy vehicle types are:

P_T – 10%
P_B – 4%
P_R – 2%

The terrain is level.

Solution

SF = $V \div$ PHF
    = 1200 ÷ 0.8
    = 1500

Correction factors:

f_d = 0.94 (80/20 distributional split)

Contd

Example 4.5 Contd

If we assume that level of service E is provided in this case, the correctional factors for both lane width and heavy vehicles can be estimated:

$f_w = 0.97$ (3.65 m wide lanes, 1.22 m (4 ft) clearance on hard shoulders)

Level terrain, level of service E, therefore $E_T = 2.0$, $E_B = 1.6$ and $E_R = 1.6$

Therefore:

$$f_{HV} = \frac{1}{1 + \{0.10(1) + 0.04(0.6) + 0.02(0.6)\}} = 0.88$$

In order to calculate the actual level of service:

$$\frac{v}{c} = \text{Service flow} \div (2800 \times f_d \times f_w \times f_{HV})$$
$$= 1500 \div (2800 \times 0.94 \times 0.97 \times 0.88)$$
$$= 0.65$$

Therefore the assumption of level of service E was a correct one and consistent with the correction factors computed.

4.4.5   *Sizing a road using the Highway Capacity Manual approach*

When sizing a new roadway, a desired level of service is chosen by the designer. This value is then used in conjunction with a design traffic volume in order to select an appropriate cross-section for the highway.

Within the US it is common practice to use a peak hour between the tenth and fiftieth highest volume hour of the year, with the thirtieth highest the most widely used. In order to derive this value for a highway, the annual average daily traffic (AADT) for the highway is multiplied by a term called the K factor, such that:

Design hourly volume (DHV) = $K_i \times \text{AADT}$ \hspace{2cm} (4.23)

where the value of K_i corresponds to the ith highest annual hourly volume

Typical values for a roadway are:

K_1 (the value corresponding to the highest hourly volume) = 0.15
K_{30} (the value corresponding to the thirtieth highest hourly volume) = 0.12

If the K_1 value is used, the road will never operate at greater than capacity but will have substantial periods when it operates at well under capacity, thus representing to some extent a waste of the resources spent constructing it. It is therefore more usual to use the K_{30} figure. While it implies that the road will be over capacity 29 hours per year, it constitutes a better utilisation of economic resources.

Finally, since DHV is a two-directional flow (as is AADT), the flow in the peak direction (the directional design hour volume (DDHV)) is estimated by multiplying it by a directional factor D:

$$DDHV = K \times D \times AADT \tag{4.24}$$

Example 4.6
A divided rural multi-lane highway is required to cope with an AADT of 40 000 vehicles per day.

A 70 mph design speed is chosen with lanes a standard 3.65 m wide and there are no obstructions within 1.83 m of any travelled edge. The traffic is assumed to be composed entirely of private cars and the driver population is ideal.

The peak hour factor is 0.9 and the directional factor, D, is estimated at 0.6. The highway is required to maintain level of service C. It is to be designed to cope with the thirtieth highest hourly volume during the year.

Calculate the required physical extent of the highway, i.e. the number of lanes required in each direction.

Solution

$$DDHV = K \times D \times AADT$$
$$= 0.12 \times 0.6 \times 40000$$
$$= 2880 \ v/h$$

We can now calculate the service flow, knowing the hourly volume and the peak hour factor, as follows:

$$SF = V \div PHF$$
$$= 2880 \div 0.9$$
$$= 3200 \ v/h$$

In order to calculate the required number of lanes, Equation 4.18 can be rearranged as follows:

$$N = SF_{(i)} \div \left(C_j \times \left(\frac{v}{c}\right)_i \times f_w \times f_{hv} \times f_p \times f_E \right) \tag{4.25}$$

Assuming the following:

C_j = 2000 (design speed = 70 mph)
f_w = 1.0 (all lanes standard, no obstructions)
f_{HV} = 1.0 (no trucks, buses or recreational vehicles in traffic stream)
f_p = 1.0 (ideal driver population)
f_E = 1.0 (rural divided multi-lane highway).

Contd

> **Example 4.6 Contd**
>
> From Table 4.2, the maximum ratio of flow to capacity for LOS C is 0.71. The number of lanes required can thus be estimated as:
>
> $$N = 3200 \div (2000 \times 0.71 \times 1.0 \times 1.0 \times 1.0 \times 1.0) = 2.25$$
>
> Therefore, rounding up, three lanes are required in each direction.

4.5   The UK approach for rural roads

4.5.1   Introduction

In the past, predicted flows in the fifteenth year after opening were used as a reference for the selection of a carriageway standard most likely to be operationally acceptable (DoT, 1985). The most recent advice note (DoT, 1997a) now advocates using opening year flows as a starting point for assessing new rural trunk road links.

Under the British system, the process for the selection of carriageway standard can be summarised as:

- Determine the 24-hour annual average daily traffic (AADT) estimate (high and low growth values) for the opening year of the scheme. These estimates must allow for the effects of induced traffic levels.
- Compare the quantities of flow derived with the ranges of flow specified for a number of carriageway standards (see Table 4.12).
- Choose those carriageway standards within those value ranges where either or both of the above flow predictions lie. For example, forecasted AADT values within the range of 25 000 to 27 000 show that D2AP and D3AP carriageway standards are suitable for assessment.
- Having taken account of any potentially important local factors such as the cost of construction and maintenance, network effects or severe impacts on the environment, both economic and environmental assessments can be carried out in order to select the optimal standard.

The process is substantially at variance with the 'level of service' approach.

Table 4.12 illustrates the opening day AADT ranges for the different carriageway widths and road types for rural highways.

4.5.2   Estimation of AADT for a rural road in its year of opening

Selection of the appropriate layout for a rural road requires that the high and low forecast for the opening year is estimated. This is undertaken using the

Table 4.12 Flow ranges for different classes of rural highway

Carriageway type	AADT range (Opening year)	Quality of access
Single 2-lane (2 × 3.65 m) (S2)	Up to 13 000	Restricted access, concentration of turning movements
Wide single carriageway (2 × 5.0 m) (WS2)	6000–21 000	Restricted access, concentration of turning movements
Dual 2-lane all-purpose carriageways (D2AP)	11 000–39 000	Restricted access, concentration of turning movements
Dual 3-lane all-purpose carriageways (D3AP)	23 000–54 000	Severe restriction of access, left turn only, clearway
Dual 2-lane motorway (D2M)	Up to 41 000	Motorway restrictions apply
Dual 3-lane motorway (D3M)	25 000–67 000	Motorway restrictions apply
Dual 4-lane motorway (D4M)	52 000–90 000	Motorway restrictions apply

factors supplied in Table 4.13 that are used in conjunction with present day AADT figures to produce future flows. These are then matched with the appropriate AADT range in Table 4.12 in order to choose the right carriageway type.

Traffic growth factors were issued in the COBA Manual (DoT, 1997b). This document contained traffic growth profiles over the years 1981 to 1996 along with forecasts of low and high traffic profiles for each year from 1997 to 2031 expressed as annual percentage growth rates.

Traffic was split into the following five categories:

- Cars (including taxis, mini-buses and camper vans)
- Light goods vehicles (LGVs) – All car type delivery vans (excluding those with twin rear tyres)
- Other goods vehicles (OGV 1) – All goods vehicles with either two axles with twin rear tyres, three axles rigid, or miscellaneous vehicles such as ambulances and tractors
- Other goods vehicles (OGV 2) – All goods vehicles with either three axles articulated or four or more axles (rigid or articulated)
- Buses and coaches (PSV) – Buses and mini-buses with the capacity for more than six passengers.

Table 4.13 below gives traffic growth rates from 1997 to 2031 for the above five categories of vehicle contained in the 1997 UK forecast. The document assumes zero growth post 2031. Values are in annual percentage growth rates rather than expressed as an index.

The average percentage of each vehicle type within the UK during 1994 was as shown in Table 4.14.

	2002–2006		2007–2011		2012–2016	
	Low	High	Low	High	Low	High
Cars	1.31	1.95	1.11	1.76	1.01	1.67
LGV	1.93	2.58	1.82	2.48	1.92	2.59
OGV 1	0.43	1.07	0.46	1.10	0.55	1.20
OGV 2	2.21	2.85	2.09	2.74	2.15	2.82
PSV	0.34	0.98	0.34	0.99	0.41	1.07
	2017–2021		2022–2026		2027–2031	
	Low	High	Low	High	Low	High
Cars	0.64	1.31	0.30	0.98	0.28	0.97
LGV	1.83	2.51	1.67	2.36	1.49	2.19
OGV 1	0.57	1.24	0.60	1.29	0.60	1.30
OGV 2	2.01	2.69	1.89	2.58	1.73	2.44
PSV	0.50	1.17	0.59	1.27	0.66	1.36

Table 4.13 Low and high growth forecasts for five different categories of road traffic (DoT, 1997b)

Cars	LGV	OGV1	OGV2	PSV
0.83	0.09	0.04	0.03	0.01

Table 4.14 Percentage of different vehicle classes in UK for 1994 (DoT, 1996)

Example 4.7

A highway is due to open on the first day of 2006. On 1 January 2002 the traffic flow was measured at an equivalent AADT of 15 000. On the basis of the forecasted AADT range in the opening year, select the appropriate rural road classification.

Assuming that the vehicular traffic is composed of the above five categories (car, LGV, OGV1, OGV2, PSV) calculate the AADT for the year after opening (2006) using the high and low growth estimates.

Solution

15 000 vehicles:

- 12 450 cars
- 1350 LGV
- 600 OGV1
- 450 OGV2
- 150 PSV

Cars:

$$\text{High growth } 2002–2006 = 12\,450 \times (1 + 0.0195)^4 = 13\,450$$
$$\text{Low growth } 2002–2006 = 12\,450 \times (1 + 0.0131)^4 = 13\,115$$

Contd

Example 4.7 Contd

LGV:

High growth 2002–2006 $= 1350 \times (1 + 0.0258)^4 = 1495$
Low growth 2002–2006 $= 1350 \times (1 + 0.0193)^4 = 1457$

OGV1:

High growth 2002–2006 $= 600 \times (1 + 0.0107)^4 = 626$
Low growth 2002–2006 $= 600 \times (1 + 0.0043)^4 = 610$

OGV2:

High growth 2002–2006 $= 450 \times (1 + 0.0285)^4 = 504$
Low growth 2002–2006 $= 450 \times (1 + 0.0221)^4 = 491$

PSV:

High growth 2002–2006 $= 150 \times (1 + 0.0098)^4 = 156$
Low growth 2002–2006 $= 150 \times (1 + 0.0034)^4 = 152$

Therefore:

AADT_{2006} (high forecast) $= 13\,450 + 1495 + 625 + 505 + 155 = 16\,231$
AADT_{2006} (low forecast) $= 13\,115 + 1457 + 610 + 491 + 152 = 15\,825$

(Giving a high growth rate over the four years of 8.2% and a low rate of 5.5%)

Table 4.12 shows that a wide single carriageway (2 × 5.0 m) with a range of 10 000 to 18 000 is the most appropriate standard. The high forecast brings the flows just over the upper limit for a standard 7.3 m wide single carriageway road (maximum capacity = 16 000)

4.6   The UK approach for urban roads

4.6.1   Introduction

The advice note TA 79/99 *Traffic Capacity of Urban Roads* (DoT, 1999) gives the maximum hourly flows for different highway categories. The maximum flows can be used as starting points in the design and evaluation of new proposed urban links. These capacities can also be utilised as a guide for assessing the adequacy of existing urban highways and the effect that any changes to their basic features (such as carriageway width) will have on the safe and efficient operation of the highway. It should be noted that, as with the assessment of rural links, both economic and environmental considerations must be taken into consideration before a final decision is made.

For the purposes of assigning capacity of different urban road types, five different highway types are considered:

- Motorways (UM)
- Urban all-purpose road Type 1 (UAP1)
- Urban all-purpose road Type 2 (UAP2)
- Urban all-purpose road Type 3 (UAP3)
- Urban all-purpose road Type 4 (UAP4).

For motorways, the prime determinant of capacity is the carriageway width as access is severely restricted, with minimal impediments to through traffic. In the case of all-purpose roads, however, flow is affected by the speed limit, the frequency of side roads, the degree of parking and loading, the frequency of at-grade pedestrian crossings and bus stops/accesses. In particular, UAP3 and UAP4 may carry high percentages of local traffic, resulting in an escalation in the levels of turning movements both at the junctions and accesses.

The design of rural roads is as detailed in Tables 4.15, 4.16, 4.17 and 4.18. Tables 4.15 and 4.16 relate carriageway widths to peak-hour flows for 2-way roads. Table 4.17 details capacities for 1-way urban roads. Table 4.18 details adjustments required to be made to the peak-hour flows where the proportion of heavy goods vehicles in the traffic stream exceeds 15%.

The flows supplied in the tables are the maximum that typical urban highways can carry consistently in one hour. It should be noted that, in most situations, the design flows on existing non-dual carriageway routes will be governed by the capacity of the terminal junction. The design flows themselves are only appropriate when the highway in question is used solely as a traffic link. Caution should therefore be exercised to ensure that the carriageway widths chosen do not correspond to maximum design flows which exceed the capacity of the junctions at peak times.

Flows for single carriageways as indicated in Table 4.15 are based on a 60/40 directional split in the flow. The flows for highway type UM as given in Table 4.16 apply to motorways where junctions are relatively closely spaced.

4.6.2   Forecast flows on urban roads

The flow ranges supplied in Tables 4.15 to 4.17 provide a guide for the assessment of appropriate carriageway standards, applicable both to new urban highways and to the upgrading of existing facilities. The purpose of these standards is to aid the highway planner in deciding on a carriageway width which will deliver an appropriate level of service to the motorists. The capacities referred to in these tables apply to highway links. The effects of junctions are dealt with in the next chapter.

If an existing route is being improved/upgraded, existing traffic flows should be measured by manual counts, with allowances made for hourly, daily or

Table 4.15 Design traffic flows for single carriageway urban roads (2-way)

Road type	Number of lanes	C'way width (m)	Peak-hour flow (veh/h)
UAP1	2	6.10	1020
High standard single	2	6.75	1320
carriageway road carrying	2	7.30	1590
predominantly through traffic	2	9.00	1860
with limited access.	2/3	10.00	2010
	3	12.30	2550
	3/4	13.50	2800
	4	14.60	3050
	4+	18.00	3300
UAP2	2	6.10	1020
Good standard single carriageway	2	6.75	1260
road with frontage access and	2	7.30	1470
more than two side roads per km	2	9.00	1550
	2/3	10.00	1650
	3	12.30	1700
	3/4	13.50	1900
	4	14.60	2100
	4+	18.00	2700
UAP3	2	6.10	900
Single carriageway road of variable	2	6.75	1100
standard with frontage access,	2	7.30	1300
side roads, bus stops and	2	9.00	1530
pedestrian crossing	2/3	10.00	1620
UAP4	2	6.10	750
Busy high street	2	6.75	900
carrying mostly local traffic with	2	7.30	1140
frontage activity (loading/	2	9.00	1320
unloading included)	2/3	10.00	1410

monthly variations in flow. If available, continuous automatic traffic count data may help identify periods of maximum flow.

In the case of proposed new highway schemes, the carriageway standard chosen by means of Tables 4.15 to 4.17 should not be used as a design tool in isolation; factors other than peak flows should be considered. Economic and environmental factors must also be considered before a final decision is taken.

4.7   Expansion of 12 and 16-hour traffic counts into AADT flows

In order to estimate the AADT for a given highway, it is not necessary to carry out a traffic count over the entire 365-day period. Use of a count of limited time duration will necessitate taking into account seasonal flow factors in order to derive an AADT valuation. In particular, certain factors for expanding 12 and 16-hour counts to values of AADT are dependent on the type of roadway and

Table 4.16 Design traffic flows for dual carriageway urban roads (2-way)

Road type	Number of lanes	C'way width (m)	Peak-hour flow (vehicles/hour)
Urban motorway	2	7.30	4000
Through route with grade-	3	11.00	5600
separated junctions, hard	4	14.60	7200
shoulders and standard			
motorway restrictions			
UAP1	2	6.75	3350
High standard dual carriageway	2	7.30	3600
road carrying predominantly	3	11.00	5200
through traffic with limited			
access.			
UAP2	2	6.75	2950
Good standard dual carriageway	2	7.30	3200
road with frontage access and	3	11.00	4800
more than two side roads per km			
UAP3	2	6.75	2300
Dual carriageway road of variable	2	7.30	2600
standard with frontage access,	3	11.00	3300
side roads, bus stops and			
pedestrian crossing			

Table 4.17 Design traffic flows for urban roads (1-way)

Road type	Number of lanes	C'way width (m)	Peak-hour flow (vehicles/hour)
UAP1	2	6.75	2950
As above	2	7.30	3250
	2/3	9.00	3950
	2/3	10.00	4450
	3	11.00	4800
UAP2	2	6.10	1800
As above	2	6.75	2000
	2	7.30	2200
	2/3	9.00	2850
	2/3	10.00	3250
	3	11.00	3550

Table 4.18 Corrections to be applied to 1 and 2-way design flows due to heavy vehicle percentages in excess of 15%

Road type	Heavy goods vehicles (%)	Total reduction on flow (vehicles/hour)
UM and UAP dual carriageway (per lane)	15–20	100
	20–25	150
Single carriageway UAP greater than 10m wide (per carriageway)	15–20	100
	20–25	150
Single carriageway UAP less than 10m wide	15–20	150
	20–25	225

the month within which the limited survey is collated, and are based on an extensive range of surveys undertaken over the past 40 years by the Transport Research Laboratory.

E-factors are used to transform 12-hour flows into 16-hour flows. M-factors are utilised to transform 16-hour flows into AADT values. These factors are supplied in Tables 4.20 and 4.21 (DoT, 1996).

For the purposes of this conversion process, highways are categorised as:

- Motorways
- Built-up trunk roads (speed limit ≤40 mph)
- Built-up principal roads (speed limit ≤40 mph)
- Non built-up trunk roads (speed limit >40 mph)
- Non built-up principal roads (speed limit >40 mph).

The above network classification is utilised to call default valuations for seasonality index (SI) as well as for the E-factors. SI is defined as the ratio of the average August weekday flow to the average weekday flow in the so-called neutral months (April, May, June, September and October). The 'neutral' months are deemed so because they are seen as being the most representative months available for extrapolation into full-year figures.

As seen in Table 4.19, the SI value defines the M-factor (a value that can be overwritten if local information is available). A good estimate of SI is derived by comparison of the weekday (Monday to Friday) flows of three-week continuous traffic counts from the month of August with those of late May/June/October.

General ranges and typical default values of seasonality index for the different classes of highway are shown in Table 4.19.

Network classification	Range of SI	Typical SI valuation
Motorway (MWY)	0.95–1.35	1.06
Built-up trunk roads (TBU)	0.95–1.10	1.00
Built-up principal roads (PBU)	0.95–1.15	1.00
Non-built-up trunk roads (TNB)	1.00–1.50	1.10
Non-built-up principal roads (PNB)	1.00–1.40	1.10

Table 4.19 Default SI values for each highway classification (DoT, 1996)

At a minimum, a 12-hour count (0700 to 1900) must be available. It should preferably be for one of the five neutral months, as traffic models not based on these are considered less reliable in terms of establishing annual flows. In the absence of local information, default E-factors for converting the 12-hour counts to 16-hour equivalents (0600 to 2200) are as shown in Table 4.20.

Network classification	E-factor
Motorway	1.15
Built-up trunk roads	1.15
Built-up principal roads	1.15
Non-built-up trunk roads	1.15
Non-built-up principal roads	1.15

Table 4.20 Default E-factor values (in the absence of local information) (DoT, 1996)

The 16-hour flow is converted to an AADT value by applying the M-factor. These valuations vary depending on the month in which the traffic count was taken and by seasonality index. In the absence of robust long-term and locally derived traffic data, default values of the M-factor for different values of SI for each of the neutral months are calculated using the following formula:

$$M = a + (b \times SI) \qquad (4.26)$$

Values of the parameters a and b for all five neutral months together with the M-factor values for differing values of seasonality index are given in Table 4.21.

Month	Parameter		M-factor		
	a	b	SI = 1.0	SI = 1.25	SI = 1.5
April	287	73	360	378	397
May	316	33	349	357	367
June	408	−57	351	337	323
September	445	−102	343	318	292
October	297	61	358	373	389

Table 4.21 M-factor scores for different values of SI (DoT, 1996)

Table 4.22 shows the M-factor values for the five neutral months for the five road classifications, given their typical SI values as indicated in Table 4.19.

Month	M-factor		
	TBU/PBU SI = 1.0	MWY SI = 1.06	TNB/PNB SI = 1.10
April	360	364	367
May	349	351	352
June	351	348	345
September	343	337	333
October	358	362	364

Table 4.22 M-factor scores for motorways, trunk roads (built-up and non-built-up) and principal roads (built-up and non-built-up) for the five neutral months

---

Example 4.8

A 12-hour traffic count carried out on the route of a proposed motorway in April established a volume of 45 000 vehicles.

What is the AADT figure?

Solution

Use E-factor to convert 12-hour value to a 16-hour value

$45\,000 \times 1.15 = 51\,750$

Use M-factor to convert 16-hour figure to AADT

For a motorway, with the traffic count taken in April, the M-factor is 364

Therefore,

Annual traffic = $51\,750 \times 364$
Annual average daily traffic = $51\,750 \times 364 \div 365 = 51\,608$ vehicles.

This flow falls within the range of a 3-lane dual motorway (D3M).

4.8   Concluding comments

The British and American methods offer contrasting methodologies for sizing a roadway. The UK system yields a recommended design range which will offer the designer a choice of types and widths. The final choice of option is made by considering economic factors such as user travel and accident costs within a cost-benefit framework. Environmental effects will also form part of the evaluation. The US system, on the other hand, relates design flows to the expected level of service to be supplied by the highway. Level of service is dependent on factors such as traffic speed and density.

4.9   References

Department of Transport (1985) *Traffic flows and carriageway width assessments*. Departmental Standard TD 20/85. The Stationery Office, London, UK.

Department of Transport (1996) Traffic Input to COBA. *Design Manual for Roads and Bridges, Volume 13: Economic Assessment of Roads*. The Stationery Office, London, UK.

Department of Transport (1997a) *Traffic Flow Ranges for Use in the Assessment of New Rural Road*. Departmental Advice Note TA 46/97. The Stationery Office, London, UK.

Department of Transport (1997b) The COBA Manual. *Design Manual for Roads and Bridges, Volume 13: Economic Assessment of Roads*. The Stationery Office, London, UK.

Department of Transport (1999) *Traffic Capacity for Urban Roads.* Departmental Advice Note TA 79/99. The Stationery Office, UK.

Greenberg, H. (1959) Analysis of traffic flow. *Operations Research*, Volume 7, Number 1.

Greenshields, B.D. (1934) A study of traffic capacity. *Proceedings of the 14th Annual Meeting of the Highway Research Board.* Highway Research Board, Washington, District of Columbia, USA.

Pipes, L.A. (1967) Car following diagrams and the fundamental diagram of road traffic. *Transport Research*, **1**, (1).

TRB (1985) *Highway Capacity Manual.* Special Report 209. Transportation Research Board, Washington, District of Columbia, USA.

Underwood, R.T. (1961) Speed, volume and density relationships. In *Quality and Theory of Traffic Flow.* Bureau of Highway Traffic, Yale, USA.

Werner, A. & Morrall, J.F. (1976) Passenger car equivalences of trucks, buses and recreational vehicles for two-lane rural highways. *Transportation Research Record* 615.

Chapter 5

The Design of Highway Intersections

5.1 Introduction

A highway intersection is required to control conflicting and merging streams of traffic so that delay is minimised. This is achieved through choice of geometric parameters that control and regulate the vehicle paths through the intersection. These determine priority so that all movements take place with safety.

The three main types of junction dealt with in this chapter are:

- Priority intersections, either simple T-junctions, staggered T-junctions or crossroads
- Signalised intersections
- Roundabouts.

All aim to provide vehicle drivers with a road layout that will minimise confusion. The need for flexibility dictates the choice of most suitable junction type. The selection process requires the economic, environmental and operational effects of each proposed option to be evaluated.

The assessment process requires the determination of the projected traffic flow at the location in question, termed the design reference flow. The range within which this figure falls will indicate a junction design which is both economically and operationally efficient rather than one where there is either gross over or underprovision. Different combinations of turning movements should be tested in order to check the performance characteristics of each junction option under consideration.

The starting point for any junction design is thus the determination of the volume of traffic incident on it together with the various turning, merging and conflicting movements involved. The basis for the design will be the flow estimate for some point in the future – the design reference flow (DRF). It is an hourly flow rate. Anything from the highest annual hourly flow to the fiftieth highest hourly flow can be used. For urban roads, use of the thirtieth highest flow is usual, with the fiftieth highest used on interurban routes.

Use of these figures implies that, during the design year in question, it can be anticipated that the design reference flow at the junction will be exceeded and a certain level of congestion experienced. If, however, the highest hourly flow was utilised, while no overcapacity will be experienced, the junction will operate at well below its capacity for a large proportion of the time, thus making such a design economically undesirable with its scale also having possible negative environmental effects due to the intrusion resulting from its sheer scale. (If the junction is already in existence then the DRF can be determined by manual counts noting both the composition of the traffic and all turning movements.)

5.2   Deriving design reference flows from baseline traffic figures

5.2.1   Existing junctions

At existing junctions, it will be possible to directly estimate peak hour and daily traffic flows together with all turning movements. In order for the measurements to be as representative as possible of general peak flow levels, it is desirable to take them during a normal weekday (Monday to Thursday) within a neutral month (April, May, June, September or October). Factoring up the observed morning and evening peak hour flows using indices given in the National Road Traffic Forecasts can lead to the derivation of DRFs for the design year (normally 15 years after opening). Flow patterns and turning proportions observed in the base year can be extrapolated in order to predict future patterns of movement.

5.2.2   New junctions

In a situation where a junction is being designed for a new road or where flow patterns through an existing junction are predicted to change significantly because of changes to the general network, flows must be derived by use of a traffic modelling process which will generate estimates of 12, 16 or 24 hour link flows for a future chosen design year. AADT flows are then obtained by factoring the 12, 16 or 24 hour flows. The AAHT is then calculated (AADT ÷ 24) and then factored to represent the appropriate highest hourly flow using derived factors. Tidal flow is then taken into consideration; generally a 60/40 split in favour of the peak hour direction is assumed. Turning proportions are also guesstimated so that the junction can be designed.

5.2.3   Short-term variations in flow

Traffic does not usually arrive at a junction at a uniform or constant rate. During certain periods, traffic may arrive at a rate higher than the DRF, at other periods lower. If the junction analysis for a priority junction/roundabout is being done with the aid of one of the Transport Research Laboratory's computer programs (PICADY/ARCADY), such variations can be allowed for using a 'flow profile'. A typical profile could involve the inputting of peak-time flows in 15-minute intervals. When calculations are being completed by hand for a priority junction/roundabout, such short-term variations may be taken into consideration by utilising an hourly flow equal to 1.125 times the DRF.

In the case of a priority junction, this adjustment should be applied to the design flows on both the minor and major arms. In the case of a roundabout intersection, this factored flow will impact not only on the entry flows to the roundabout but also the circulating flows within the intersection.

5.2.4   Conversion of AADT to highest hourly flows

Appendix D14 of the *Traffic Appraisal Manual* (DoT, 1996) originally detailed factors linking AADT to the tenth, thirtieth, fiftieth, hundredth and two hundredth highest annual hourly peak flow for three classes of roads:

- Main urban
- Interurban
- Recreational interurban.

Particularly on urban highways, where peaks are less marked, the thirtieth highest flow may be most appropriate. On an interurban route, the fiftieth highest might apply. On recreational routes where peaks occur infrequently, the two hundredth highest may be the value most consistent with economic viability. The general implication is that where the design flow is exceeded some degree of congestion will result, but this is preferable and economically more justifiable to the situation where congestion will never occur and the road is under capacity at all times.

The values originally given in Table 5A of Appendix D14 of the *Traffic Appraisal Manual* range between approximately 2.4 and 4.4. The use of these national expansion figures for converting AADT to peak hour flows is no longer recommended. Rather it is advised that local traffic data be used in order to compile such factors.

5.3   Major/minor priority intersections

5.3.1   Introduction

A priority intersection occurs between two roads, one termed the 'major' road and the other the 'minor' road. The major road is the one assigned a permanent priority of traffic movement over that of the minor road. The minor road must give priority to the major road with traffic from it only entering the major road when appropriate gaps appear. The principal advantage of this type of junction is that the traffic on the major route is not delayed.

The principle at the basis of the design of priority intersections is that it should reflect the pattern of movement of the traffic. The heaviest traffic flows should be afforded the easiest paths. Visibility, particularly for traffic exiting the minor junction, is a crucial factor in the layout of a priority intersection. Low visibility can increase the rate of occurrence of serious accidents as well as reducing the basic capacity of the intersection itself.

Priority intersections can be in the form of simple T-junctions, staggered junctions or crossroads, though the last form should be avoided where possible as drivers exiting the minor road can misunderstand the traffic priorities. This may lead to increased accidents.

Diagrammatic representations of the three forms are given in Fig. 5.1.

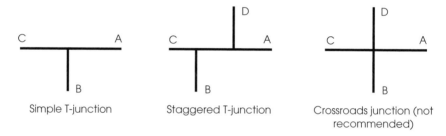

Simple T-junction Staggered T-junction Crossroads junction (not recommended)

Figure 5.1 Three forms of priority intersection.

Within the two main junction configurations mentioned above (T-junction/staggered junction), there are three distinct types of geometric layout for a single-carriageway priority intersection:

- *Simple junctions*: A T-junction or staggered junction without any ghost or physical islands in the major road and without channelling islands in the minor road approach (see Fig. 5.2).

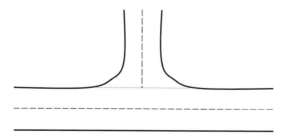

Figure 5.2 Simple T-junction.

- *Ghost island junctions*: Usually a T-junction or staggered junction within which an area is marked on the carriageway, shaped and located so as to direct traffic movement (see Fig. 5.3).

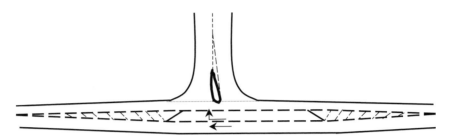

Figure 5.3 Ghost island junction.

- *Single lane dualling*: Usually a T-junction or staggered junction within which central reservation islands are shaped and located so as to direct traffic movement (see Fig. 5.4).

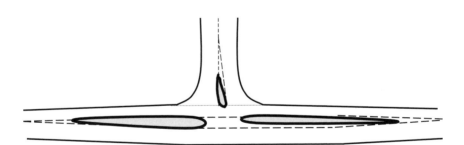

Figure 5.4 Single lane dualling.

The type or level of junction utilised depends on the flows from both the major and minor roads. The 'simple' layout is appropriate for new junctions in rural locations where the 2-way AADT on the minor road is not expected to exceed 300 vehicles with the major road 2-way AADT not exceeding 13 000 vehicles. For single carriageway roads, the different levels of T-junctions appropriate to a range of flow combinations are illustrated in Figure 5.5 (DoT, 1981). The information takes into account geometric and traffic delays, entry and turning capacities and accident costs.

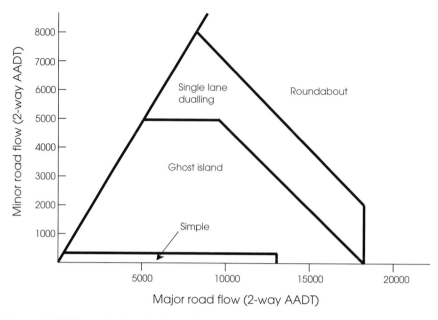

Figure 5.5 Different levels of priority intersections for various major and minor road flows in the design year.

Example 5.1 – Determination of flows at a priority intersection
An existing simple major/minor priority junction at the intersection of two single carriageway 2-lane roads is to be upgraded. The major route is a major interurban highway. Figure 5.6 shows the annual average daily traffic (AADT) flow ranges predicted for each of the arms to the intersection in the design year, assumed in this case to be 15 years after opening.

Taking the upper AADT flows of 10250 for the major carriageway and 3000 on the minor carriageway, Fig. 5.5 indicates that the junction must

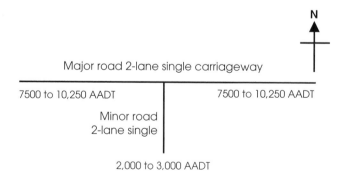

Figure 5.6 AADT 2-way flows.

Contd

Example 5.1 Contd

be upgraded to a ghost island junction. Using these AADT values, the 2-way annual average hourly traffic (AAHT) on all three arms can be computed by dividing each AADT value by 24, giving the results shown in Fig. 5.7.

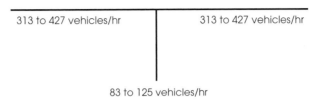

Figure 5.7 AAHT flows.

It is decided on economic grounds to design the junction to cater for the thirtieth highest annual hourly flow. On the basis of local traffic count data, it is found that the AAHT flow on each of the approach roads is translated into a design peak hour flow by multiplying by a factor of 3.0, with the results shown in Fig. 5.8.

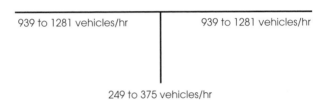

Figure 5.8 Design peak hour flows (2-way).

Directional splits derived from local traffic data are applied to these design flows in order to obtain the final directional flows, in this case a 70:30 split in favour of flows from the southern and eastern approaches, as shown in Fig. 5.9.

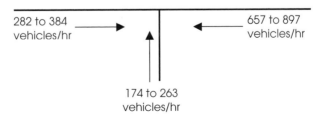

Figure 5.9 Directional flows from all arms.

Contd

Example 5.1 Contd

Turning movements were assessed based on present day traffic flow data from the junction, which indicated the proportions in Fig. 5.10.

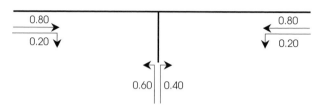

Figure 5.10 Turning movements.

Taking the upper flow ranges from Fig. 5.9, combined with the turning proportions indicated in Fig. 5.10, a set of flows is derived which can then be used to analyse the junction, as shown in Fig. 5.11.

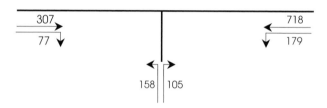

Figure 5.11 Final upper range flows (vehicles per hour).

5.3.2   *Equations for determining capacities and delays*

Once the flows to be analysed have been determined and a generally appropriate geometric layout has been settled on, it is then necessary to establish the capacity of each traffic movement through the priority junction. In each case, it is mainly dependent on two factors:

- The quantity of traffic in the conflicting and merging traffic movements
- General geometric properties of the junction.

The traffic flows relevant to the determination of these capacities are shown in Fig. 5.12.

The following equations for predicting the capacity of turning traffic streams are contained in TA 23/81 (DoT, 1981), the advice note on the design of major/minor intersections:

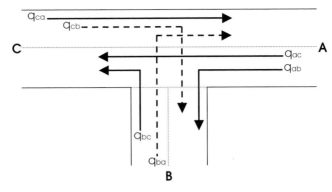

Figure 5.12 Relevant flows for determining capacity of major movements through priority intersections.

$$\mu_{ba} = D(627 + 14W_{cr} - Y[0.364q_{ac} + 0.114q_{ab} + 0.229q_{ca} + 0.520q_{cb}]) \qquad (5.1)$$

$$\mu_{bc} = E(745 - Y[0.364q_{ac} + 0.114q_{ab}]) \qquad (5.2)$$

$$\mu_{cb} = F(745 - 0.364Y[q_{ac} + q_{ab}]) \qquad (5.3)$$

In these equations the stream capacities and flows are measured in passenger car units per hour.

The geometric characteristics of the junction, represented in Equations 5.1 to 5.3 by D, E and F and specific to each opposed traffic stream, are represented as:

$$D = [1 + 0.094(w_{ba} - 3.65)][1 + 0.0009(Vr_{ba} - 120)][1 + 0.0006(Vl_{ba} - 150)]$$
$$E = [1 + 0.094(w_{bc} - 3.65)][1 + 0.0009(Vr_{bc} - 120)]$$
$$F = [1 + 0.094(w_{cb} - 3.65)][1 + 0.0009(Vr_{cb} - 120)]$$

Parameters:

μ = Stream capacity

q_{ab} = Measured flow of stream *a-b*

q_{ac} = Measured flow of stream *a-c*

q_{ca} = Measured flow of stream *c-a*

q_{cb} = Measured flow of stream *c-b*

w_{ba} = Average lane width over a distance of 20 m available to waiting vehicles in the stream *b-a*, metres

w_{bc} = Average lane width over a distance of 20 m available to waiting vehicles in the stream *b-c*, metres

w_{cb} = Average lane width available to waiting vehicles in the stream *c-b*, metres

Vr_{ba} = Visibility to the right from a point 10 m back from the give-way line for vehicles making the *b-a* manoeuvre, metres

Vl_{ba} = Visibility to the left from a point 10 m back from the give-way line for vehicles making the *b-a* manoeuvre, metres

Vr_{bc} = Visibility to the right from a point 10 m back from the give-way line for vehicles making the *b-c* manoeuvre, metres

Vr_{cb} = Visibility to the right, along the major road for traffic crossing traffic performing the *c-b* manoeuvre, metres

W_{cr} = Width of the central reserve (only for dual carriageways), metres

Y = $(1 - 0.0345\,W)$

W = Total major road carriageway width, metres.

Similar equations exist for estimating capacities at staggered junctions and crossroads (see Semmens, 1985).

The determination of queue lengths and delays is of central importance to assessing the adequacy of a junction. When actual entry flows are less than capacity, delays and queue sizes can be forecast using the steady state approach. With this method, as demand reaches capacity, delays and therefore queue lengths tend towards infinity.

The steady state result for the average queue length L is:

$$L = \rho + C\rho^2/(1 - \rho) \tag{5.4}$$

where
C is a constant depending on the arrival and service patterns; for regular vehicle arrivals $C = 0$, for random arrivals $C = 1$. In the interests of simplicity, the latter case is assumed.

$$\rho = \text{flow } (\lambda) \div \text{capacity } (\mu)$$

Therefore, Equation 5.4 can be simplified as:

$$L = \rho/(1 - \rho) \tag{5.5}$$

Thus, as $\rho \to 1$, $L \to \infty$

In reality, this is not the case with queue lengths where the ratio of flow to capacity reaches unity. Thus, at or near capacity, steady state theory overestimates delays/queues.

On the other hand, within deterministic theory, the number of vehicles delayed depends on the difference between capacity and demand. It does not take into account the statistical nature of vehicle arrivals and departures and seriously underestimates delay, setting it at zero when demand is less than or equal to capacity.

The deterministic result for the queue length L after a time t, assuming no waiting vehicles, is:

$$L = (\rho - 1)\mu t \tag{5.6}$$

Thus, as $\rho \to 1$, $L \to 0$

At a busy junction, demand may frequently approach capacity and even exceed it for short periods. Kimber and Hollis (1979) proposed a combination of Equations 5.5 and 5.6, using co-ordinate transformations to derive the following equations for average queue length and delay.

Queue length

$$L = 0.5 \times ((A^2 + B)^{1/2} - A)\tag{5.7}$$

where

$$A = \frac{(1-\rho)(\mu t)^2 + (1-L_0)\mu t - 2(1-C)(L_0 + \rho\mu t)}{\mu t + (1-C)}\tag{5.8}$$

$$B = \frac{4(L_0 + \rho\mu t)\lfloor \mu t - (1-C)(L_0 + \rho\mu t)\rfloor}{\mu t + (1-C)}\tag{5.9}$$

Delay per unit time

Kimber and Hollis (1979) evaluated the relevant area under the queue length curve in order to derive an expression for delay.

$$D_t = 0.5 \times ((F^2 + G)^{1/2} - F)\tag{5.10}$$

where

$$F = \frac{(1-\rho)(\mu t)^2 - 2(L_0 - 1)\mu t - 4(1-C)(L_0 + \rho\mu t)}{2(\mu t + 2(1-C))}\tag{5.11}$$

$$G = \frac{2(2L_0 + \rho\mu t)\lfloor \mu t - (1-C)(2L_0 + \rho\mu t)\rfloor}{\mu t + 2(1-C)}\tag{5.12}$$

Delay per arriving vehicle

Kimber and Hollis (1979) derived the following expression as a measure of the average delay per arriving vehicle over an interval. The expression has two parts; the first relates to those suffered in the queue while the second $(1/\mu)$ relates to those encountered at the give-way line.

$$D_v = 0.5 \times ((P^2 + Q)^{1/2} - P) + 1/\mu\tag{5.13}$$

where

$$P = [0.5 \times (1-\rho)t] - \frac{1}{\mu}(L_0 - C)\tag{5.14}$$

$$Q = \frac{2Ct}{\mu}\left(\rho + \frac{2L_0}{\mu t}\right)\tag{5.15}$$

Parameters:

μ = capacity

ρ = ratio of actual flow to capacity = q/μ

q = actual flow

C = 1 for random arrivals and service patterns

C = 0 for regular arrivals and service patterns

L_0 = queue length at start of time interval under consideration

t = time.

It is reasonable to set C equal to 1. Therefore, Equations 5.8, 5.9, 5.11 and 5.12 can be simplified as:

$$A = (1-\rho)\mu t + 1 - L_0 \tag{5.16}$$

$$B = 4(L_0 + \rho\mu t) \tag{5.17}$$

$$F = [0.5 \times (1-\rho)\mu t] - L_0 + 1 \tag{5.18}$$

$$G = 4L_0 + 2\rho\mu t \tag{5.19}$$

$$P = [0.5 \times (1-\rho)t] - \frac{1}{\mu}(L_0 - 1) \tag{5.20}$$

$$Q = \frac{2t}{\mu}\left(\rho + \frac{2L_0}{\mu t}\right) \tag{5.21}$$

Example 5.2 – Computing capacities, queue lengths and delays at a priority intersection

Figure 5.13 indicates a set of design reference flows for the evening peak hour at a proposed urban-based priority intersection. All flows are in passenger car units per hour.

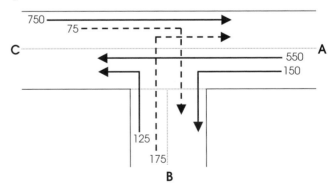

Figure 5.13 Design reference flows at priority intersection.

Contd

Example 5.2 Contd

Estimate the ratio of flow to capacity for each of the opposed movements and, for the one with the highest ratio, estimate the average queue length and delay per vehicle during the peak hour.

Geometric parameters:

The width of the major carriageway (W) is 9.5 m.

The lane widths for traffic exiting the minor road turning both left and right (w_{ba}, w_{bc}) are 2.5 m.

The lane width on the major road for traffic waiting to turn right onto the minor road (w_{cb}), is 2.2 m.

Visibility to the right and left for traffic exiting the minor road turning right (Vr_{ba}, Vl_{ba}) is 30 m and 50 m respectively.

Visibility to the right for traffic exiting the minor road turning left (Vr_{bc}) is 30 m.

Visibility to the right for traffic exiting the major road, turning right onto the minor road (Vr_{cb}) is 50 m.

Solution

When using Equations 5.1 to 5.3 for estimating the ratios of flow to capacity for the various opposed turning movements, the design reference flows shown in Fig. 5.13 should be multiplied by 1.125 in order to allow for short-term variations in the traffic flow. (These short-term variations are allowed for in PICADY.)

Factored flows:

q_{ab} = 169 (150 × 1.125)

q_{ac} = 619 (550 × 1.125)

q_{ca} = 844 (750 × 1.125)

q_{cb} = 84 (75 × 1.125)

q_{ba} = 197 (175 × 1.125)

q_{bc} = 141 (125 × 1.125)

Contd

Example 5.2 Contd

Parameters D, E and F:

$$D = [1 + 0.094(2.5 - 3.65)][1 + 0.0009(30 - 120)][1 + 0.0006(50 - 150)] = 0.77$$
$$E = [1 + 0.094(2.5 - 3.65)][1 + 0.0009(30 - 120)] = 0.82$$
$$F = [1 + 0.094(2.2 - 36.5)][1 + 0.0009(50 - 120)] = 0.81$$
$$Y = [(1 - (0.345 \times 9.5))] = 0.67$$

Capacities:

$$\mu_{ba} = 0.77\{627 - 0.67[0.364(619) + 0.114(169) + 0.229(844) + 0.520(84)]\}$$
$$= 234 \, \text{pcu/hr}$$
$$\mu_{bc} = 0.82\{745 - 0.67[(0.364(619)) + 0.114(169)]\}$$
$$= 477 \, \text{pcu/hr}$$
$$\mu_{cb} = 0.81[745 - (0.364 \times 0.67)(619 + 169)]$$
$$= 448 \, \text{pcu/hr}$$

Ratios of flow to capacity (RFC):

For the most critical movement (traffic exiting onto the major road, turning right) the ratio is just below the maximum allowed in urban areas of 0.85. The maximum reduces to 0.75 in rural areas.

Table 5.1 RFCs for opposed flow movements

Stream	Flow (pcu/hr)	Capacity (pcu/hr)	RFC
q_{ba}	197	234	0.84
q_{bc}	141	477	0.30
q_{cb}	84	448	0.19

Queue length at stream B-A:

$$L = 0.5 \times ((A^2 + B)^{1/2} - A)$$

Assuming random arrivals (therefore C = 1)

$$A = (1 - \rho)\mu t + 1 - L_0$$
$$B = 4(L_0 + \rho\mu t)$$
$$\rho = 0.84$$
$$\mu = 234 \, \text{pcu/hr}$$
$$t = 1 \, \text{hour}$$
$$L_0 = 0 \, \text{(no cars waiting at start of time period)}$$

Contd

Example 5.2 Contd

Therefore

$$A = [(1 - 0.84) \times 234 \times 1] + 1$$
$$= 38.44$$
$$B = 4 \times 0.84 \times 234 \times 1$$
$$= 786.24$$
$$L = 0.5 \times ((38.44^2 + 786.24)^{1/2} - 32.45)$$
$$= 5 \, cars$$

Total delay during peak hour

$$D_t = 0.5 \times (F^2 + G)^{1/2} - F)$$
$$F = [0.5 \times (1 - 0.84) \times 234 \times 1] + 1$$
$$= 11.72$$
$$G = 2 \times 0.84 \times 234 \times 1$$
$$= 393.12$$
$$D_t = 0.5 \times ((19.72^2 + 393.12)^{1/2} - 17.72)$$
$$= 4.1 \, hours.$$

Delay per arriving vehicle (excluding delay at stop-line)

$$D_v = 0.5 \times ((P^2 + Q)^{1/2} - P) + 1/\mu$$
$$P = [0.5 \times 0.16] + \frac{1}{234}$$
$$= 0.08 + 0.00427$$
$$= 0.0843$$
$$Q = \frac{2}{234} \times 0.84$$
$$= 0.0072$$
$$D_v = \{0.5 \times [(0.0843^2 + 0.0072)^{1/2} - 0.0843]\} + 1/234$$
$$= 0.0219 \, hours$$
$$= 79 \, seconds \, per \, vehicle$$

Thus, the average delay per vehicle is 1.3 minutes, with a queue length of five vehicles.

5.3.3   *Geometric layout details* (DoT, 1995)

Horizontal alignment

In the ideal situation, the priority intersection should not be sited where the major road is on a sharp curve. Where this is unavoidable, it is preferable that the T-junction is located with the minor junction on the outside of the curve.

Vertical alignment

The preferred location for a priority intersection is on level terrain, or where the gradient of the approach roads does not exceed an uphill or downhill gradient of 2%. Downhill gradients greater than this figure induce excessively high speeds, while uphill approaches prevent the drivers from appreciating the layout of the junction.

Visibility

Along the major and minor roads, approaching traffic should be able to see the minor road entry from a distance equal to the desirable minimum sight stopping distance (SSD). The required SSD depends on the chosen design speed and varies from 70 m for a design speed of 50 km/h to 295 m for 120 km/h. (Further details on sight stopping distances are given in Chapter 6.)

In addition, on the minor road, from a distance x metres back along the centre line of the road, measured from a continuation of the line of the nearside edge of the running carriageway of the major road, the approaching driver should be able to see clearly points on the left and right on the nearside edge of the major road running carriageway a distance y away. The x value is set at 9 m. The y value varies depending on the chosen design speed along the major road (Table 5.2).

Design speed along major road (km/h)	y distance (metres)
50	70
60	90
70	120
85	160
100	215
120	295

Table 5.2 y visibility distances from the minor road (see Fig. 5.14 for an illustration of x and y)

Dedicated lane on major road for right-turning vehicles

In the case of all non-simple junctions, provision of a right-turning lane, the lane itself should not be less than 3 m. It requires the provision of a turning length, a deceleration length and a direct taper length. Figure 5.14 illustrates these parameters for a ghost island junction. The turning length (a) is set at 10 m; the deceleration length (b) varies from 25 m to 110 m depending on the design speed on the major route. The taper (c) varies from 1 in 20 to 1 in 30 for ghost island/single lane dualling intersections. (Greater values are required for dual carriageway intersections.)

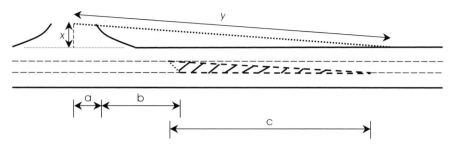

Figure 5.14 Illustration of turning, deceleration and taper lengths for a ghost island junction.

5.4   Roundabout intersections

5.4.1   Introduction

In order to control merging and conflicting traffic flows at an intersection, a roundabout performs the following two major functions:

- It defines the priority between traffic streams entering the junction, usually on the basis that traffic wanting to join the circulatory flow must give way to the traffic to their right already circulating in the roundabout. (In the UK and Ireland traffic circulates in a clockwise direction.)
- It causes the diversion of traffic from its preferred straight-line path, requiring drivers to slow down as they enter the junction.

In order to work efficiently, sufficient gaps must appear in the circulating flows on the roundabout that drivers then accept. Traffic on the entry arms can thus enter, circulate and then leave at their desired exit arm. Its operation has, therefore, certain similarities to that of a priority intersection and, as we shall see later in the chapter, the design procedure in both has certain similarities. The situation is more complex in the case of a roundabout intersection as there is no clear identifiable major road traffic flow that can be used as a basis for designing the junction, with the circulating flow depending on the operation of all entry arms to it.

If properly designed, the angles at which traffic merges/diverges will be small. This, combined with the relatively slow traffic speeds on the roundabout, will help reduce accident rates.

As seen from Fig. 5.5, when traffic flows at an intersection are relatively low, adequate control can be attained using the priority option. As flow levels increase, however, with this intersection type, delays/queue lengths become excessive and some alternative form is required. While grade-separated junctions may be the preferred option at high flow levels, the expense involved may be prohibitive. For this reason, particularly in an urban setting, at-grade roundabouts or signalised intersections become viable junction options at levels of flow above those suitable for priority control. If the cost of land is an important factor, traffic signals will be preferred as land requirements for a standard 3 or 4-arm conventional roundabout would be greater. However, right-turning vehicles can cause operational difficulties at signal-controlled junctions, particularly where volumes within this phase are large.

Roundabouts have difficulty dealing with unbalanced flows, in which case signalisation may be preferable. In situations where flows are relatively well balanced, and where three or four entry arms exist, roundabouts cope efficiently with the movement of traffic. Where the number of arms exceeds four, however, efficiency may be affected by the failure of drivers to understand the junction layout. It may also prove difficult to correct this even with comprehensive direction signing.

In addition to their ability to resolve conflicts in traffic as efficiently as possible, roundabouts are often used in situations where there is:

- A significant change in road classification/type

- A major alteration in the direction of the road

- A change from an urban to a rural environment.

5.4.2   Types of roundabout

Mini-roundabout (Fig. 5.15)

Mini-roundabouts can be extremely successful in improving existing urban junctions where side road delay and safety are a concern. Drivers must be made aware in good time that they are approaching a roundabout. Mini-roundabouts consist of a 1-way circulatory carriageway around a reflectorised, flush/slightly raised circular island less than 4m in diameter which can be overrun with ease by the wheels of heavy vehicles. It should be domed to a maximum height of 125mm at the centre for a 4m diameter island, with the height reduced pro rata for smaller islands. The approach arms may or may not be flared. Mini-roundabouts are used predominantly in urban areas with speed limits not

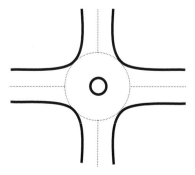

Figure 5.15 Mini-roundabout configuration.

exceeding 48 km/h (50 mph) (DoT, 1993). They are never used on highways with high speed limits. In situations where physical deflection of vehicle paths to the left may be difficult to achieve, road markings should be employed in order to induce some vehicle deflection/speed reduction. If sufficient vehicle deflection cannot be achieved, the speed of the traffic on the approach roads can be reduced using traffic calming techniques.

Because of the short distance between the entry points to the roundabout, drivers arriving at the intersection must monitor very closely the movements of other vehicles both within the junction and on the approaches in order to be in a position to react very quickly when a gap occurs.

Normal roundabout (Fig. 5.16)

A normal roundabout is defined as a roundabout having a 1-way circulatory carriageway around a kerbed central island at least 4 m in diameter, with an

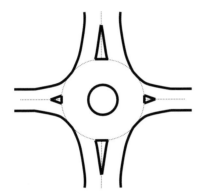

Figure 5.16 Normal roundabout configuration.

inscribed circle diameter (ICD) of at least 28 m and with flared approaches to allow for multiple vehicle entry. The number of recommended entry arms is either three or four. If the number is above four, the roundabout becomes larger with the probability that higher circulatory speeds will be generated. In such situations double roundabouts may provide a solution (DoT, 1993).

Double roundabout (Fig. 5.17)

A double roundabout can be defined as an individual junction with two normal/mini-roundabouts either contiguous or connected by a central link road or kerbed island. It may be appropriate in the following circumstances:

- For improving an existing staggered junction where it avoids the need to realign one of the approach roads
- In order to join two parallel routes separated by a watercourse, railway or motorway

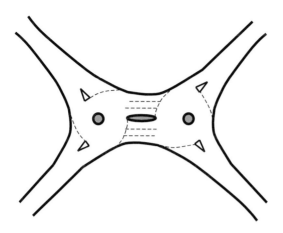

Figure 5.17 Double roundabout configuration (with central link road).

- At existing crossroads intersections where opposing right-turning movements can be separated
- Catering for junctions with more than four entries and overloaded single roundabouts where overall capacity can be increased by reducing the circulating flow travelling past critically important entry points.

In situations where the double roundabout is composed of two mini-roundabouts, the speed limit on the approaches must not exceed 48 km/h (30 mph).

Other forms

Other roundabout configurations include *two-bridge* roundabouts, *dumbbell* roundabouts, ring junctions and *signalised* roundabouts.

Example 5.3 – Determination of flows at a roundabout intersection

A junction is to be constructed at the intersection of a single carriageway 2-lane road (deemed the minor road) and a 2-lane dual carriageway (the major road). Figure 5.18 shows the annual average daily traffic (AADT) flow ranges predicted for each of the arms to the intersection in the design year, assumed in this case to be 12 years after opening.

Taking the upper AADT flows of 20000 for the major carriageway and 8000 for the minor carriageway, Fig. 5.5 indicates that the most appropriate form of junction is a roundabout. Using the AADT values, the average 2-way annual average hourly traffic (AAHT) on all four approach roads can be computed by dividing each AADT value by 24, with the results shown in Fig. 5.19.

Contd

Example 5.3 Contd

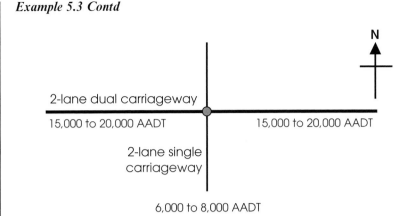

Figure 5.18 AADT flows for major and minor roads.

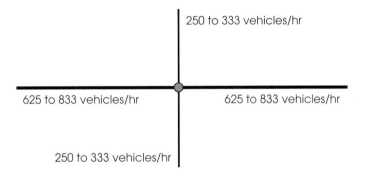

Figure 5.19 AAHT 2-way flows.

As with Example 5.1, it is decided on economic grounds to design the junction to cater for the thirtieth highest annual hourly flow. On the basis of local traffic count data, it is found that the AAHT flow on each of the approach roads is translated into a design peak hour flow by multiplying by a factor of 3.0, with the results shown in Fig. 5.20.

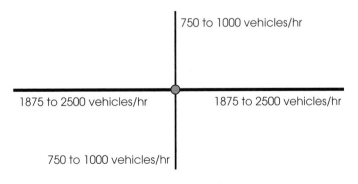

Figure 5.20 Design peak hour flows (2-way).

Contd

Example 5.3 Contd

Directional splits derived from local traffic data are applied to these design flows in order to obtain the final directional flows, in this case, a 60:40 split in favour of entry flows from the southern and western approaches, as shown in Fig. 5.21.

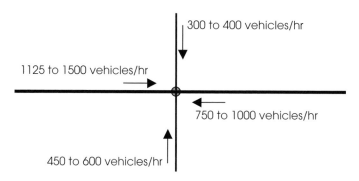

300 to 400 vehicles/hr

1125 to 1500 vehicles/hr

750 to 1000 vehicles/hr

450 to 600 vehicles/hr

Figure 5.21 Directional flows.

Using information on local traffic flow patterns, turning proportions were deduced as shown in Fig. 5.22.

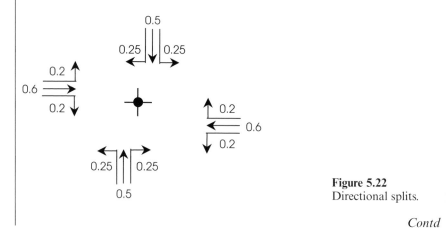

Figure 5.22
Directional splits.

Contd

Example 5.3 Contd

Using these splits, and taking the upper flow ranges from Fig. 5.21, a set of design reference flows is derived which can then be used to analyse the roundabout intersection (Fig. 5.23).

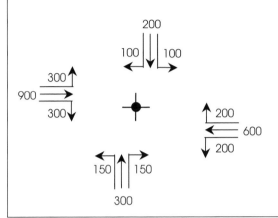

Figure 5.23 Final upper range flows (vehicles per hour).

5.4.3   *Traffic capacity at roundabouts*

The procedure for predicting the capacity of roundabouts is contained in TA 23/81 (DoT, 1981), based on research done at the Transport Research Laboratory (Kimber, 1980). The design reference flows are derived from forecast traffic levels as explained above. The capacity itself depends mainly on the capacities of the individual entry arms. This parameter is defined as entry capacity and itself depends on geometric features such as the entry width, approach half-width, entry angle and flare length. The main geometric parameters relevant to each entry arm of a roundabout are illustrated in Fig. 5.24.

The predictive equation for entry capacity (Q_E) used with all forms of mini-roundabouts or normal at-grade roundabouts is:

Q_E = entry capacity into circulatory area (vehicles per hour)
   $= k(F - fcQc)$ (5.22)

where
Qc = flow in the circulatory area in conflict with the entry (vehicles per hour)
k $= 1 - 0.00347(\phi - 30) - 0.978 [(1/r) - 0.05]$
F $= 303X_2$
fc $= 0.21tD (1 + 0.2 X_2)$
tD $= 1 + 0.5/(1 + M)$
M $= \exp [(D - 60)/10]$
X_2 $= v + (e - v)/(1 + 2S)$
S $= 1.6(e - v)/l'$

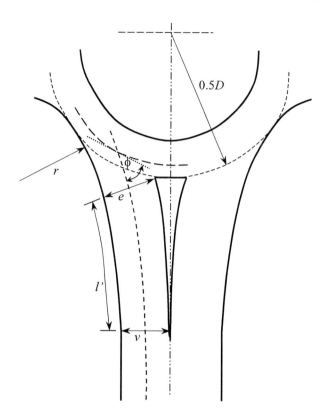

Figure 5.24
Geometric
parameters of a
roundabout.

and

e = entry width (metres) – measured from a point normal to the near kerbside

v = approach half-width – measured along a normal from a point in the approach stream from any entry flare

l' = average effective flare length – measured along a line drawn at right angles from the widest point of the entry flare

S = sharpness of flare – indicates the rate at which extra width is developed within the entry flare

D = inscribed circle diameter – the biggest circle that can be inscribed within the junction

ϕ = entry angle – measures the conflict angle between entering and circulating traffic

r = entry radius – indicates the radius of curvature of the nearside kerbline on entry.

Equation 5.22 applies to all roundabout types except those incorporating grade-separated junctions. Where this is the case, the F term is multiplied by 1.1 and the fc term by 1.4, i.e.:

$$Q_{E(\text{grade sep})} = k(1.1F - 1.4(fcQc))$$ (5.23)

The maximum ranges together with those recommended for design are as shown in Table 5.3.

Table 5.3 Geometric parameters for roundabouts

Symbol	Description	Allowable range	Recommended range
e	Entry width	3.6–16.5 m	4.0–15.0 m
v	Approach half-width	1.9–12.5 m	2.0–7.3 m
l'	Average flare length	1 m to infinity	1–100 m
S	Flare sharpness	Zero–2.9 m	—
D	Inscribed circle diameter	13.5–171.6 m	15–100 m
ϕ	Entry angle	0–77°	10–60°
r	Entry radius	3.4 m to infinity	6–100 m

Design reference flow

When analysing the capacity of a roundabout intersection, the capacity of each of the entry arms is assessed and compared with the traffic flow expected at peak hours within the design year. This ratio of flow to capacity (RFC) for each traffic movement, in the same manner as for priority junctions, directly indicates whether the roundabout will operate efficiently in the chosen design year.

TA 23/81 states that if an entry RFC of 70% occurs, queuing will be avoided in 39 out of 40 peak hours. A maximum RFC of 0.85 is recommended as this will result in an intersection whose provision is economically justified yet will not cause excessive delay and disruption.

Again, as stated earlier, use of the manual procedure will require that all design reference flows (DRF) are multiplied by 1.125 in order to allow for short-term variations in traffic flow at the roundabout within the peak hour.

Example 5.4 – Assessing the ratio of flow to capacity for each of the entry points to a 3-arm roundabout

Figure 5.25 indicates a set of design reference flows for the evening peak hour at a proposed roundabout intersection. All flows are in vehicles per hour.

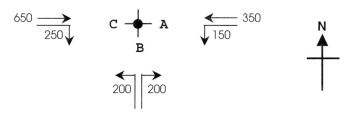

Figure 5.25 Design reference flows.

Contd

Example 5.4 Contd

Estimate the ratio of flow to capacity for each entry point. A heavy goods vehicle content of 10% is assumed.

The following are the geometric parameters assumed for each entry arm at the junction:

e = 7.5 m
v = 4 m (east and west arms), 3.65 m (south arm)
l' = 10 m
D = 28 m
ϕ = 30°
r = 10 m

Therefore

k = 0.951
S = 0.56 (east and west arms), 0.62 (south arm)
X_2 = 5.65 (east and west arms), 5.37 (south arm)
F = 1712 (east and west arms), 1629 (south arm)
M = 0.0408
tD = 1.48
fc = 0.662 (east and west arms), 0.645 (south arm)

For east arm (Arm A):

$$Q_E = k(F - fcQc)$$
$$= 0.9511(1712 - 0.6622Qc)$$
$$= 1629 - 0.63Qc$$

For south arm (Arm B):

$$Q_E = k(F - fcQc)$$
$$= 0.951(1629 - 0.65Qc)$$
$$= 1549 - 0.614Qc$$

For west arm (Arm C):

$$Q_E = k(F - fcQc)$$
$$= 0.9511(1712 - 0.662Qc)$$
$$= 1629 - 0.63Qc$$

Values of entry and circulating flows

In order to estimate the circulating flow and thus the entry capacity of each entry movement using the manual procedure, the entry flow values from Fig. 5.25 are multiplied by a factor of 1.125. The derived flows are then multiplied by 1.1 in order to convert the vehicles to passenger car units (pcu) (Table 5.4).

Table 5.4 Entry flows

	ARM A		ARM B		ARM C	
	q_{AB}	q_{AC}	q_{BA}	q_{BC}	q_{CA}	q_{CB}
Flow in veh/hr	150	350	200	200	650	250
Flow in pcu/hr (×1.1)	165	385	220	220	715	275
Factored flow (×1.125)	186	433	248	248	804	310
Factored entry flow (pcu/hr)	619		495		1114	
Factored circulating flow (pcu/hr)	310		433		248	

The fully factored flows entering, exiting and circulating within the round-about are illustrated diagrammatically in Fig. 5.26 and graphically in Fig. 5.27.

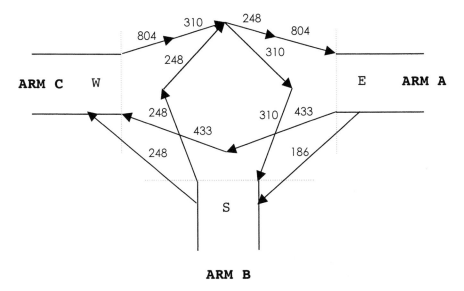

Figure 5.26 Factored entry, circulating and exiting flows (pcu/hr).

Entry capacities

Given the circulating flows estimated above, the entry capacity for each of the three arms can be computed as:

Arm A
$$Q_E = 1629 - 0.63Qc$$
$$= 1629 - (0.63 \times 310)$$
$$= 1433 \, \text{pcu/hr}$$

Arm B
$$Q_E = 1549 - 0.614Qc$$
$$= 1549 - (0.614 \times 433)$$
$$= 1283 \, \text{pcu/hr}$$

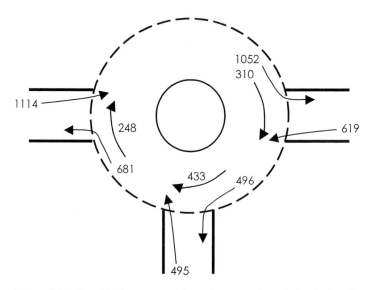

Figure 5.27 Graphical representation of entry, exit and circulating flows.

Arm C
$$Q_E = 1629 - 0.63Q_c$$
$$= 1629 - (0.63 \times 248)$$
$$= 1473 \, \text{pcu/hr}$$

The ratios of flow to capacity for each entry arm are given in Table 5.5.

Table 5.5 RFCs for each entry arm

Stream	Flow (pcu/hr)	Capacity (pcu/hr)	RFC
Arm A	619	1433	0.43
Arm B	496	1283	0.39
Arm C	1114	1473	0.76

All movements are below the maximum allowed RFC of 0.85. Arm A is slightly in excess of 70% so some queuing will occur during peak travel times. Delays will however not be significant.

5.4.4   Geometric details

The geometric guidelines for roundabout junctions are set out in TD 16/93 (DoT, 1993). While it is not proposed to go into detail on the geometric considerations that must be addressed when designing a roundabout, the following are brief notes from TD 16/93 regarding design of the following main parameters:

- Entry width
- Entry angle
- Entry radius
- Entry deflection
- Inscribed circle diameter (ICD)
- Circulatory carriageway
- Main central island.

Entry width

TD16/93 states that it is good practice to add at least an extra lane to the lanes on the entry approach. (This has been done at all entry arms in the above example. In all cases one lane has been widened to two at the entry point.)

Entry angle

It is recommended that the entry angle should be between 10 and 60°. (In the above example, an entry angle of 30° was used at all entry points.)

Entry radius

The absolute minimum radius should be 6 m and, in order to cater for heavy goods vehicles, it should preferably not be less than 10 m.

Entry deflection

This is one of the main determinants of safety on a roundabout, indicating the deflection to the left imposed on all entering vehicles. For each entry arm, the tightest radius of the entry path curvature, measured over a distance of at least 20–25 m, should not exceed 100 m.

Inscribed circle diameter (ICD)

In order to accommodate the turning movement of a standard 15.5 m long articulated vehicle, the inscribed circle diameter must be at least 28 m. The ICD in the above example is exactly 28 m.

Circulatory carriageway

The width of the circulatory should not exceed 15 m and should lie between 1.0 and 1.2 times the maximum entry width. The chosen design sets the circulatory carriageway at 9 m, approximately 1.2 times the maximum entry width value (7.6 m).

Main central island

The main central island will have a radius of 2 m, the maximum allowable for an ICD of 28 m, again to allow the safe movement through the junction of large articulated vehicles. In order to limit the circulatory carriageway, therefore, a low profile subsidiary central island should be provided, extending the radius of the central island out to a total of 5 m. The subsidiary island provides deflection for standard vehicles while allowing overrun by the rear wheels of articulated vehicles.

5.5   Basics of traffic signal control: optimisation and delays

5.5.1   Introduction

Traffic signals work on the basis of allocating separate time periods to conflicting traffic movements at a highway intersection so that the available carriageway space is utilised as efficiently and safely as possible. Priority can be varied with time through the cycle of the signals. Within urban areas in particular, in situations where a road has a number of intersections along its entire length, signal linking can be used as a method for allowing almost continuous progression of traffic through the route.

In the UK the decision to install traffic lights at a given junction is arrived at through assessing in economic terms the reduction in delay resulting from it together with the forecasted improvement in accident characteristics and placing these against the computed capital and operating costs. In the UK the usual sequence for traffic signals is red, red/amber, green and amber. In Ireland, the sequence is red, green and amber.

The installation of traffic signals is justified by the need to:

- Reduce delay to motorists and pedestrians moving through the junction
- Reduce accidents at the junction
- Improve the control of traffic flow into and through the junction in particular and the area in general, thereby minimising journey times
- Impose certain chosen traffic management policies.

Traffic signals can however have certain disadvantages:

- They must undergo frequent maintenance along with frequent monitoring to ensure their maximum effectiveness
- There can be inefficiencies during off-peak times leading to increases in delay and disruption during these periods
- Increases in rear-end collisions can result
- Signal breakdown due to mechanical/electrical failure can cause serious interruption in traffic flow.

As with priority junctions and roundabouts, the basis of design in this instance is the ratio of demand to capacity for each flow path. In addition, however, the setting of the traffic signal in question is also a relevant parameter. The capacity of a given flow path is expressed as its saturation flow, defined as the maximum traffic flow capable of crossing the stop line assuming 100% green time. Traffic signals will normally operate on the basis of a fixed-time sequence, though vehicle actuated signals can also be installed. The fixed-time sequence may be programmed to vary depending on the time of day. Typically, a separate programme may operate for the morning and evening peaks, daytime off-peak and late night/early morning conditions.

5.5.2   *Phasing at a signalised intersection*

Phasing allows conflicting traffic streams to be separated. A phase is characterised as a sequence of conditions applied to one or more traffic streams. During one given cycle, all traffic within a phase will receive identical and simultaneous signal indications.

Take, for example, a crossroads intersection, where north-south traffic will conflict with that travelling from east to west. Since the number of conflicts is two, this is the number of required phases. With other more complex intersections, more than two phases will be required. A typical example of this is a crossroads with a high proportion of right-turning traffic on one of the entry roads, with movement necessitating a dedicated phase. A three-phase system is therefore designed. Because, as will be shown further below, there are time delays associated with each phase within a traffic cycle, efficiency and safety dictate that the number be kept to a minimum.

The control of traffic movement at a signalised junction is often described in terms of the sequential steps in which the control at the intersection is varied. This is called stage control, with a stage usually commencing from the start of an amber period and ending at the start of the following stage. Stages within a traffic cycle are arranged in a predetermined sequence. Figures 5.28 and 5.29 illustrate, respectively, typical two-phase and three-phase signalised intersections.

5.5.3   *Saturation flow*

The capacity for each approach to an intersection can be estimated by summing the saturation flows of the individual lanes within each of the pathways. The main parameters relevant to the estimation of capacity for each approach are:

- *Number, width and location of lanes*
  The overall capacity of an approach road is equal to the sum of the saturation flows for all lanes within it. It is assumed that the average lane width

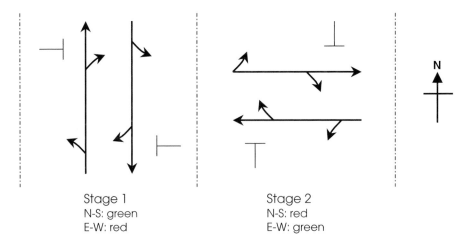

Stage 1
N-S: green
E-W: red

Stage 2
N-S: red
E-W: green

Figure 5.28 Typical stages within a two-phase system.

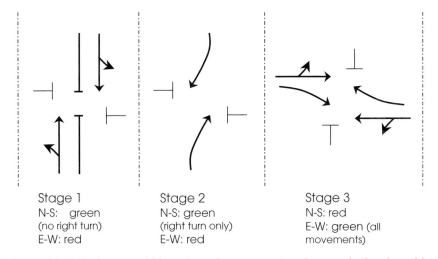

Stage 1
N-S: green
(no right turn)
E-W: red

Stage 2
N-S: green
(right turn only)
E-W: red

Stage 3
N-S: red
E-W: green (all
movements)

Figure 5.29 Typical stages within a three-phase system (two lanes each direction with a dedicated lane for right-turning vehicles in all cases).

on the approach to a signalised intersection is 3.25 m. At this value, assuming zero gradient, the saturation flow for nearside lanes is set at 1940 pcu/hr, increasing to 2080 pcu/hr for non-nearside lanes.

● *Weather conditions*
In wet weather conditions, saturation flows decrease to 6% below their dry weather values.

● *Gradient*
For every 1% increase in uphill gradient (measured over a 60 m distance back from the stop-line) the saturation flow value will decrease by 2%. Downhill gradients do not affect saturation values

- *Turning movements*

  In situations where the turning traffic is unopposed, saturation flows will decrease as the proportion of turning traffic increases. Where turning traffic is opposed, the saturation flow will depend on the number of gaps in the opposing traffic flow together with the amount of storage space available to those vehicles making this traffic movement.

The equations derived by Kimber *et al.* (1986) for predicting saturation flow are:

Unopposed traffic streams

The saturation flow is given by:

$$S_1 = (S_0 - 140d_n)/(1 + 1.5f/r) \text{ pcu/hr} \tag{5.24}$$

where

$$S_0 = 2080 - 42d_g \times G + 100(w - 3.25) \tag{5.25}$$

and

d_n = 1 (nearside lane) or 0 (non-nearside lane)

f = proportion of turning vehicles in the lane under scrutiny

r = radius of curvature of vehicle path, metres

d_g = 1 (uphill entry roads) or 0 (downhill entry roads)

G = percentage gradient of entry road

w = entry road lane width, metres.

Opposed traffic streams

In this instance, the saturation flow in a given lane for right-turning opposed streams is given by:

$$S_2 = S_g + S_c \text{ (pcu/hr)} \tag{5.26}$$

The two components of this equation are computed as follows.

(1) S_g

S_g is the saturation flow occurring during the 'effective green' time within the lane of opposed mixed turning traffic.

$$S_g = (S_0 - 230)/(1 + (T - 1)f) \tag{5.27}$$

$$T = 1 + 1.5/r + t_1/t_2 \tag{5.28}$$

$$t_1 = 12X_0^2/(1 + 0.6(1 - f)N_s) \tag{5.29}$$

$$X_0 = q_0/\lambda n_1 s_0 \tag{5.30}$$

$$t_2 = 1 - (fX_0)^2 \tag{5.31}$$

X_0 is the degree of saturation on the opposing entry arm (ratio of flow to capacity)

q_0 is the actual flow on the opposing arm, measured in vehicles per hour of green time (excluding non-hooking right-turning vehicles)

λ is the effective green time divided by the total cycle time, C

n_1 is the number of lanes within the opposing entry arm

s_0 is the saturation flow for each of the lanes on the opposing entry arm

N_s is the number of storage spaces within the junction which the right-turning vehicles can use so as not to block the straight ahead stream.

(2) S_c

S_c is the saturation flow occurring after the 'effective green' time within the lane of opposed mixed turning traffic.

(During a traffic phase, the effective green time is the actual green time plus the amber time but minus a deduction for starting delays.)

$$S_c = P(1 + N_s) \, (f X_0)^{0.2} \times 3600/\lambda C \qquad (5.32)$$

$$P = 1 + \Sigma_i(\alpha_i - 1)p_i \qquad (5.33)$$

P is the conversion factor from vehicles to passenger car units

α_i is the pcu value of vehicle type i

p_i is the proportion of vehicles of type i.

The passenger car unit values used in connection with the design of a signalised junction are given in Table 5.6.

Table 5.6 pcu values

Vehicle type	pcu equivalent
Car/light vehicle	1.0
Medium commercial vehicle	1.5
Heavy commercial vehicle	2.3
Bus/coach	2.0

Example 5.5 – Calculation of saturation flow for both opposed and unopposed traffic lanes

The approach road shown in Fig. 5.30 is composed of two lanes, both 3.25 m wide. The nearside lane is for both left-turning and straight-ahead traffic, with a ratio of 1:4 in favour of the straight through movement. The non-nearside lane is for right-turning traffic. This movement is opposed. The degree of saturation of the opposing traffic from the north is 0.6.

Contd

Example 5.5 Contd

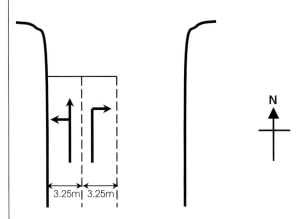

3.25m| 3.25m|

Figure 5.30 Diagram of 2-lane approach.

The turning radius for all vehicles is 15 m. Zero gradient and one 30 second effective green period per 60 second traffic cycle can be assumed ($\lambda = 0.5$). The traffic is assumed to be composed of 90% private cars and 10% heavy commercial vehicles.

Calculate the saturation flow for each of the two movements. Assume two storage spaces exist within the junction.

Solution

For the nearside lane, the saturation flow can be calculated using Equations 5.24 and 5.25:

$S_1 = (S_0 - 140d_n)/(1 + 1.5f/r)$ pcu/hr

where $S_0 = 2080 - 42d_g \times G + 100(w - 3.25)$
The values of the relevant parameters are:

$d_n = 1$
$f = 0.2$
$r = 15$
$G = 0$
$w = 3.25$

Therefore
$S_0 = 2080 - (42d_g \times 0) + 100(3.25 - 3.25)$
$= 2080$
and
$S_1 = (2080 - 140)/(1 + (1.5 \times 0.2/15))$
$= 1902$ pcu/hr

Contd

Example 5.5 Contd

For the non-nearside lane, the saturation flow can be calculated using Equations 5.26 to 5.33:

$$f = 1 \text{ (all vehicles turning)}$$
$$X_0 = 0.6$$
$$N_s = 2 \text{ (storage spaces)}$$
$$t_1 = 12 \, (0.6)^2/(1 + 0.6(1 - 1) \times 2)$$
$$= 4.32$$
$$t_2 = 1 - (1 \times 0.6)^2$$
$$= 0.64$$
$$T = 1 + (1.5/15) + (4.32/0.64)$$
$$= 1 + 0.1 + 6.75$$
$$= 7.85$$
$$S_g = (2080 - 230)/(1 + (7.85 - 1) \times 1)$$
$$= 1850/7.85$$
$$= 236 \, \text{pcu/hr}$$
$$P = 0.9 \times 1 + (0.1 \times 2.3)$$
$$= 1.13$$
$$S_c = 1.13 \times (1 + 2) \, (1 \times 0.6)^{0.2} \times 3600/(0.5 \times 60)$$
$$= 367 \, \text{pcu/hr}$$

Therefore:
$$S_2 = 236 + 367$$
$$= 603 \, \text{pcu/hr}$$

5.5.4   *Effective green time*

Effective green time is defined as the length of time that would be required to get a given discharge rate over the stop line if the flow commenced and finished simultaneously and instantaneously on the change of colour as displayed on the signal head.

An analysis of the flow of vehicles across the stop line at an intersection permits the effective green time to be estimated. The discharge of vehicles across the stop line starts at the beginning of the green period and finishes at the end of the amber period. The intervals of time between the start of actual green time and the start of effective green and between the end of effective green time and the end of the amber period are termed lost time.

At the start of any given cycle, when the light goes green and traffic begins to move off, the flow across the stop line rises from zero, gradually increasing until saturation flow is achieved. The flow level remains steady until the light turns amber at the end of the phase. Some vehicles will stop; others may take some time to do so. The flow returns to zero as the lights turn red.

From Fig. 5.31 it can been seen that the actual green time plus the amber period is equal to the effective green time plus the two periods of lost time at the beginning and end of the cycle. The effective green time is thus the length of time during which saturation flow would have to be sustained in order to obtain the same quantity of traffic through the lights as is achieved during an actual green period. It is denoted by a rectangle in Fig. 5.31. This rectangle has exactly the same area as that under the actual flow curve.

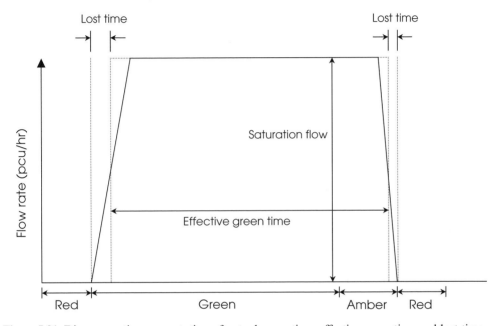

Figure 5.31 Diagrammatic representation of actual green time, effective green time and lost time.

Normally, the lost time is assumed to be taken as equal to 2 seconds, with the amber time set at 3 seconds. Effective green time is thus equal to actual green time plus 1 second.

5.5.5 *Optimum cycle time*

Assuming the signal system at an intersection is operating on fixed-time control, the cycle length will directly affect the delay to vehicles passing through the junction. There is always an element of lost time every time the signal changes. If the cycle time is short, this lost time becomes a significant proportion of it, leading to inefficiencies in the working of the junction and consequent lengthy delays. Too long a cycle will result in all queuing vehicles being cleared in the early part of the green period, with the only vehicles crossing the stop line in the latter part being those that subsequently arrive, often quite widely spaced. This too is inefficient. (The discharge of traffic through a junction is at its most efficient when there is a waiting queue on the approach road.)

Webster (1958) developed a set of formulae for establishing the optimum signal settings in order to minimise the total delay to all streams on the approach roads. The optimum cycle time C_o is obtained from:

$$C_o = (1.5L + 5) \div (1 - Y) \tag{5.34}$$

where
L = total lost time per cycle
Y = the sum of the maximum y values for all of the phases which make up the cycle
(y is the ratio of actual flow to saturation flow on each approach)

There is a minimum cycle time of 25 seconds based on safety considerations. A maximum cycle time of 120 seconds is considered good practice. Normally, the cycle time will lie within the range of 30 to 90 seconds.

Lost time

Lost time per cycle consists of the time lost during the green period (generally taken as 2 seconds per phase) plus the time lost during what is known as the *intergreen period*. The intergreen period is defined as the period between one phase losing right of way and the next phase gaining right of way, or the time between the end of green on one phase and the start of green on the next, The intergreen period provides a suitable time during which vehicles making right turns can complete their manoeuvre safely having waited in the middle of the intersection.

If the amber time during the intergreen period is 3 seconds and the total intergreen period is 5 seconds, this gives a lost time of 2 seconds, as this is the period of time for which all lights show red or red/amber, a time during which no vehicle movement is permitted. The period of time lost to traffic flow is referred to as lost time during the intergreen period. It should not be confused with lost time due to starting delays at the commencement of each phase.

Table 5.7 shows an example of a two-phase system.

Table 5.7 Typical intergreen period within a two-phase signal system

Phase 1	G	A	A	A	R	R	Red
		Amber period (3 s)			Lost time (2 s)		←5 seconds intergreen period (3 s + 2 s)
Phase 2	R	R	R	R	R/A	R/A	Green

Example 5.6 – Calculation of optimum cycle time

The actual and saturation flows for a three-phase signal system are detailed in Table 5.8. The phasing details are given in Fig. 5.32. The intergreen period is set at 5 seconds, the amber time at 3 seconds and the lost time due to starting delays at 2 seconds.

Calculate the optimum cycle time for this intersection.

Table 5.8 Actual flows, saturation flows and y valuations

Stage	Movement	Flow (pcu/h)	Saturation flow (pcu/h)	y
1	1A	460	1850	0.25*
	1B	120	600	0.20
	1C	395	1800	0.22
	1D	100	550	0.18
2	2A	475	1900	0.25*
	2B	435	1900	0.23
3	3A	405	1850	0.22*
	3B	315	1850	0.17

*Maximum values

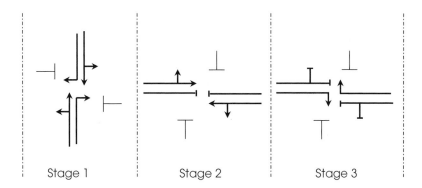

Figure 5.32 Staging diagram.

Solution

The three phases for the intersection in question, together with the movements for each of the individual approaches, are:

Stage 1

Approach 1A – north/south movement (straight ahead and left-turning)
Approach 1B – north/south movement (right-turning)
Approach 1C – south/north movement (straight ahead and left-turning)
Approach 1D – south/north movement (right-turning)

Contd

Example 5.6 Contd

Stage 2

Approach 2A – east/west movement (straight ahead and left-turning)
Approach 2B – west/east movement (straight ahead and left-turning)

Stage 3

Approach 3A – east/west movement (right-turning)
Approach 3B – west/east movement (right-turning)
For each approach, the maximum y value is marked with an asterisk in
Table 5.8.

Stage 1: $y_{max} = 0.25$
Stage 2: $y_{max} = 0.25$
Stage 3: $y_{max} = 0.22$

For each of the three phases, the lost time due to starting delays is 2 seconds,
as is the lost time during the intergreen period. The total lost time is therefore 12 seconds (4×3 seconds).
The optimum cycle time is calculated as follows:

$$C_o = (1.5L + 5) \div (1 - Y)$$

where

$$Y = \Sigma y_{max} = 0.25 + 0.25 + 0.22 = 0.72$$

Therefore

$$C_o = (1.5(12) + 5) \times (1 - 0.72)$$
$$= 23 \div 0.28$$
$$= 82 \text{ seconds}$$

5.5.6   *Average vehicle delays at the approach to a signalised intersection*

Webster (1958) derived the following equation for estimating the average delay
per vehicle at a signalised intersection:

$$d = \frac{c(1-\lambda)^2}{2(1-\lambda x)} + \frac{x^2}{2q(1-x)} - 0.65 \times \left(\frac{c}{q^2}\right)^{1/3} x^{(2+5\lambda)} \tag{5.35}$$

where
d = average delay per vehicle
c = cycle length
λ = effective green time divided by cycle time

q = flow
s = saturation flow
$x = q/\lambda s$

The first term in Equation 5.35 relates to the delay resulting from a uniform rate of vehicle arrival. The second term relates to the delay arising from the random nature of vehicle arrivals. The third term is an empirically derived correction factor, obtained from the simulation of the flow of vehicular traffic.

The above formula can be simplified as:

$$d = cA + \frac{B}{q} - C \qquad (5.36)$$

where

$$A = \frac{(1-\lambda)^2}{2(1-\lambda x)}$$

$$B = \frac{x^2}{2(1-x)}$$

C = correction term, which can be taken as 10% of the sum of the first two terms

Equation 5.35 can thus be written in approximate form as:

$$d = 0.9 \times \left(cA + \frac{B}{q} \right) \qquad (5.37)$$

Example 5.7 – Calculation of average vehicle delay at the approach to a signalised junction

An approach has an effective green time of 65 seconds and an optimum cycle time of 100 seconds. The actual flow on the approach is 1000 vehicles per hour, with its saturation flow estimated at 1750 vehicles per hour.

Calculate the average delay per vehicle using both the precise and approximate formulae.

Solution

c = 100 s
λ = 0.65
q = 1000 veh/h = 0.278 veh/s
s = 1750 veh/h = 0.486 veh/s
x = 0.278/(0.65 × 0.486) = 0.88

Contd

Example 5.7 Contd

$$A = \frac{(1-0.65)^2}{2(1-(0.65 \times 0.88))} = 0.14$$

$$B = \frac{0.88^2}{2(1-0.88)} = 3.23$$

$$C = 0.65 \times \left(\frac{100}{0.278^2}\right)^{1/3} 0.88^{5.25} = 3.6$$

Using the precise formula:

$$d = \frac{c(1-\lambda)^2}{2(1-\lambda x)} + \frac{x^2}{2q(1-x)} - 0.65 \times \left(\frac{c}{q^2}\right)^{1/3} x^{(2+5\lambda)}$$
$$= (100 \times 0.14) + (3.23 \div 0.278) - 3.6$$
$$= 14 + 12 - 4$$
$$= 22 \text{ seconds}$$

Using the approximate formula:

$$d = 0.9 \times \left(cA + \frac{B}{q}\right)$$
$$= 0.9 \times (14 + 12)$$
$$= 23 \text{ seconds}$$

5.5.7   *Average queue lengths at the approach to a signalised intersection*

It is normal practice to estimate the queue length at the beginning of the green period as this is the instant at which it will be greatest.

If the approach is assumed to be unsaturated, then whatever queue forms during the red period will be completely discharged by the end of the green period. Where this is the case, the maximum queue formed is equal to the product of the actual flow and the effective red time for the approach. (The effective red time is the cycle time minus the effective green time, i.e. the length of time for which the signal on the approach in question is effectively red.)

The formula for the unsaturated case is therefore:

$$N_u = qr \tag{5.38}$$

where

N_u = queue length at the commencement of the green period (assuming the approach is unsaturated)

q = actual flow rate

r = effective red period (cycle time – effective green time)

If it is assumed that the intersection is at a saturated level of flow, the queue length will vary gradually over any given interval of time. The average queue length can thus be estimated as the product of the actual flow, q, and the average vehicle delay, d. To this value, an estimate accounting for cyclical fluctuations caused by short-term variations in flow during the red and green periods must be added. These can range from zero to qr, therefore an average value of $qr/2$ is taken.

Combining these two terms, an expression for the saturated case is:

$$N_s = qd + 1/2\,qr \tag{5.39}$$

where

N_s = queue length at the commencement of the green period (assuming the approach is saturated)

d = average delay per vehicle on the approach (see previous section).

At a minimum, in this case, d will equal $r/2$, therefore the equation for the saturated case reduces to that for the unsaturated case, i.e. $N_s = qr$. Since this is the minimum value of N_s, Equation 5.39 can be adjusted as:

$N_s = qd + 1/2\,qr$ or qr, whichever is greater.

Example 5.8 – Calculation of average queue length at the approach to a signalised junction
Taking the figures from the previous example, calculate the average queue length at the approach.

Solution

$c = 100\,\mathrm{s}$
$q = 0.278$ veh/s
$r = 35\,\mathrm{s}$
$d = 22\,\mathrm{s}$

Therefore, taking the saturated case:

$$N_s = qd + 1/2\,qr \text{ or } qr$$
$$= (0.278 \times 22) + 1/2\,(0.278 \times 35) \text{ or } (0.278 \times 35)$$
$$= 10.98\,(\text{say } 11) \text{ or } 9.73\,(\text{say } 10)$$
$$= 11 \text{ vehicles}$$

The computer program OSCADY (Burrow, 1987; Binning, 1999) can be used to analyse new and existing signalised junctions, producing optimum cycle times, delays and queue lengths as well as information on predicted accident frequencies at the junctions under examination.

5.5.8   *Signal linkage*

Within an urban setting, where signalised intersections are relatively closely spaced, it is possible to maximise the efficiency of flow through these junctions through signal co-ordination. This can result in the avoidance of excessive queuing with consequent tailbacks from one stop line to the preceding signals, and in the ability of a significant platoon of traffic passing through the entire set of intersections without having to stop. Lack of co-ordination can result in some vehicles having to stop at each of the junctions.

One method for achieving this co-ordination is through use of a time-and-distance diagram, with time plotted on the horizontal axis and distance on the vertical axis. The rate of progression of any given vehicle through the network of signals is denoted by the slope of any line charted on it.

In order to construct the diagram, each intersection is examined individually and its optimum cycle time computed. The one with the largest required cycle time (called the 'key intersection') is identified and taken up as the cycle time for the entire network C_1. Knowing the lost time per cycle, the effective green time and hence the actual green time for the key intersection can be computed. The actual green time along the main axis of progression in this instance will determine the minimum actual green time along this axis at the other junctions within the network. With regard to the other non-key intersections, the maximum actual green time in each case is derived through determination of the smallest acceptable green time for the minor road phases:

$$\text{Min. effective green}_{\text{minor route}} = (y_{\text{side}} \times C_1) \div 0.9 \tag{5.40}$$

The minimum actual green time for the side/minor roads can then be calculated by addition of the lost time per phase and subtraction of amber time. (As stated earlier, actual time is 1 second less than effective time.) By subtracting this value from the cycle time plus the intergreen period, the maximum actual green time permitted for the junction in question along the main axis of movement can be calculated.

When these minima and maxima have been determined, the time-and-distance diagram can be plotted once the distance between the individual intersections in the network is known and an average speed of progression in both directions along the major axis is assumed.

Example 5.9 – Calculation of time-and-distance diagram
Three two-phase signalised intersections are spaced 400 m apart. The main
axis of flow is in the north/south direction. Details of the actual and saturation flows at each of the junctions are given in Table 5.9. The starting delays
are taken as 2 seconds per green period, the amber period is 3 seconds in all
cases and the period during which all lights show red during a change of
phase is 2 seconds.

Locate the critical intersection, calculate the minimum and maximum
actual green times and outline the construction of the time-distance diagram
indicating how vehicles will progress along the main axis of flow.

The junction is illustrated diagrammatically in Fig. 5.33.

Solution

Intergreen period = 10 seconds (5 seconds per phase, 3 s amber + 2 s red)
Lost time = 8 seconds in total per cycle (4 seconds per phase, 2 s starting and
            2 s during intergreen)

Firstly the optimum cycle time for each intersection must be calculated
applying Equation 5.34 to the three sets of data in Table 5.9:

$$C_o = (1.5L + 5) \div (1 - \Sigma y_{crit})$$

Values of y_{crit} for each intersection in the network are indicated in Table 5.9.

Table 5.9 Critical ratios for each intersection in the network

Intersection	Approach	Actual flow (vehicles/hour)	Saturation flow (vehicles/hour)	y	y_{crit}
A1	North	800	3200	0.250	0.375
	South	1200	3200	0.375	
	East	800	1800	0.444	0.444
	West	500	1550	0.323	
A2	North	1100	3200	0.344	0.363
	South	1160	3200	0.363	
	East	1020	2150	0.474	0.474
	West	525	1800	0.291	
A3	North	800	3200	0.250	0.313
	South	1000	3200	0.313	
	East	800	1800	0.444	0.444
	West	400	1800	0.222	

Based on the critical ratios in Table 5.9, the optimum cycle time for each
intersection can be computed as:

Contd

Example 5.9 Contd

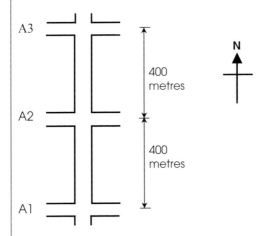

Figure 5.33
Diagrammatic layout
of network of urban-
based intersections.

Junction A1:
$$C_o = [(1.5 \times 8) + 5] \div [1 - (0.375 + 0.444)]$$
$$= 94 \text{ seconds}$$

Junction A2:
$$C_o = [(1.5 \times 8) + 5] \div [1 - (0.363 + 0.474)]$$
$$= 104 \text{ seconds}$$

Junction A3:
$$C_o = [(1.5 \times 8) + 5] \div [1 - (0.313 + 0.444)]$$
$$= 70 \text{ seconds}$$

From the above figures it can be seen that the longest of the three cycle times is for junction A2 – 104 seconds. This is the key junction and is thus adopted for the entire network.

Minimum actual green time:
Assuming a cycle time of 104 seconds, the effective green time can be estimated along the main north/south axis by use of the y_{crit} values at junction A2. Of the 104 seconds, 8 seconds is lost time, therefore the combined effective green time in both directions is 96 seconds (104 – 8). The relative values of the critical ratios are then used to estimate the proportion of this green time allocated to the north/south and east/west directions.

Since
$$y_{crit(N/S)} = 0.363$$
$$y_{crit(E/W)} = 0.474$$

Contd

Example 5.9 Contd

$$(\text{Minimum effective green time})_{\text{N/S}} = 96 \times \left[\frac{0.363}{0.363 + 0.474}\right]$$
$$= 42 \text{ seconds}$$

Given that starting delays are 2 seconds and the amber time is 3 seconds, the actual green time is equal to the effective green time minus 1 second:

$$(\text{Minimum actual green time})_{\text{N/S}} = 42 - 1$$
$$= 41 \text{ seconds}$$

This figure is adopted as the minimum green time for the remaining two junctions within the network (A1 and A3).

Maximum actual green time:
The upper limit for actual green time at A1 and A3 is computed by consideration of the minimum effective green time required by traffic at these junctions along their east/west axis (termed side traffic):

$$(\text{Minimum effective green time})_{\text{E/W}} = (C_1 \times y_{\text{crit(E/W)}}) \div 0.9$$
$$= (104 \times y_{\text{crit(E/W)}}) \div 0.9$$
$$= 116 \times y_{\text{crit(E/W)}}$$

For both A1 and A3:

$$y_{\text{crit(E/W)}} = 0.444$$

Therefore

$$(\text{Minimum effective green time})_{\text{E/W}} = 116 \times 0.444$$
$$= 52 \text{ seconds}$$

Again, given that starting delays are 2 seconds and the amber time is 3 seconds, the actual green time is equal to the effective green time minus 1 second:

$$(\text{Minimum actual green time})_{\text{E/W}} = 52 - 1 = 51 \text{ seconds}$$

The total actual green time for both the north/south and east/west phases is 94 seconds, estimated from the subtraction of the intergreen period (10 s) from the total cycle time (104 s). Subtracting the minimum actual green time along the side route from this figure yields the maximum actual green time along the main north/south axis. Therefore, for both A1 and A3 intersections:

$$(\text{Maximum actual green time})_{\text{N/S}} = 94 - 51 = 43 \text{ seconds}$$

Contd

Example 5.9 Contd

Let us assume a speed of 30 km/h which means that it takes a vehicle just under 48 seconds to progress from one junction to the next. Knowing this, together with the minimum and maximum actual green times (41 s and 43 s), the time/distance diagram illustrating the ability of vehicles to progress in both directions through the network of intersections can be compiled.

An illustration of the time-distance diagram for the above example is shown in Fig. 5.34. Note that the vehicles travel between junctions during approximately half the cycle time.

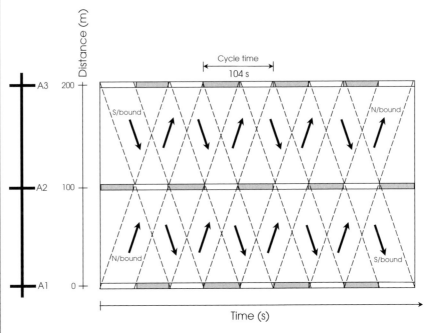

Figure 5.34 Illustration of time-distance diagram for Example 5.7.

The computer programs TRANSYT (Robertson, 1969) and SCOOT (Hunt *et al.*, 1981) can be used for signal optimisation and traffic control system analysis. TRANSYT automatically finds the optimum timing that will co-ordinate the operation of a signalised network of intersections. It produces what is termed a *performance index* for the network, assessing it on the basis of a weighted combination of delays and stops. As in the above worked example, TRANSYT assumes that all signals have a common cycle time and that the minimum green and red periods are known. SCOOT, unlike TRANSYT, is an on-line traffic control system, using measurements from road detectors in order to compile cyclic traffic flow profiles for each junction in the network being examined. These profiles are utilised to compute the best possible set of signal timings for all signals in the network area.

5.6 Concluding remarks

Within the design process for an at-grade intersection, whether it is a priority junction, roundabout or signalised junction, the same basic design procedure is employed. Firstly the traffic data must be collected, indicating the proposed loading on the junction during peak times. Data on the physical characteristics of the site must also be available, in particular horizontal and vertical alignments in the vicinity. Design standards may then dictate what type of intersection is employed. The design of the junction itself will be an iterative process, where layouts may be altered based on operating, cost and environmental concerns.

The design of grade separated intersections is not detailed within this text. See O'Flaherty (1997) for details of this intersection type.

5.7 References

Binning, J.C. (1999) *OSCADY 4 – User Guide* (Application Guide 25). Transport and Road Research Laboratory, Crowthorne, UK.

Burrow, I.J. (1987) *OSCADY – A computer program to model capacities, queues and delays at isolated traffic signal junctions*. TRRL Report RR105. Transport and Road Research Laboratory, Crowthorne, UK.

DoT (1981) Junctions and Accesses: Determination of Size of Roundabouts and Major/Minor Junctions. Departmental Advice Note TA 23/81. *Design Manual for Roads and Bridges, Volume 6: Road Geometry*. The Stationery Office, London, UK.

DoT (1985) Traffic flows and carriageway width assessments. Departmental Standard TD 20/85. *Design Manual for Roads and Bridges, Volume 5: Assessment and Preparation of Road* Schemes. The Stationery Office, London, UK.

DoT (1993) *The geometric design of roundabouts*. Departmental Standard TD 16/93. The Stationery Office, London, UK.

DoT (1995) Geometric Design of Major/Minor Priority Junctions, Departmental Standard TD 42/95. *Design Manual for Roads and Bridges, Volume 6: Road Geometry*. The Stationery Office, London, UK.

DoT (1996) Traffic Appraisal Manual. *Design Manual for Roads and Bridges, Volume 12: Traffic Appraisal of Road Schemes*. The Stationery Office, London, UK.

Hunt, P.B., Robertson, D.I. & Bretherton, R.D. (1981) *SCOOT – A traffic responsive method of coordinating signals*. Laboratory Report LR 1014. Transport and Road Research Laboratory, Crowthorne, UK.

Kimber, R.M. (1980) *The traffic capacity of roundabouts*. TRRL Report LR942. Transport and Road Research Laboratory, Crowthorne, UK.

Kimber, R.M. & Hollis, E.M. (1979) *Traffic queues and delays at road junctions.* TRRL Report LR909. Transport and Road Research Laboratory, Crowthorne, UK.

Kimber, R.M., McDonald, M. & Hounsell, N.B. (1986) *The prediction of saturation flows for road junctions controlled by traffic signals.* TRRL Research Report 67. Transport and Road Research Laboratory, Crowthorne, UK.

O'Flaherty, C.A. (1997) *Transport Planning and Traffic Engineering.* Butterworth Heinemann, Oxford.

Robertson, D.I. (1969) *TRANSYT – A traffic network study tool.* Laboratory Report LR 253. Transport Research Laboratory, Crowthorne, UK.

Semmens, M.C. (1985) *PICADY2 – An enhanced programme for modelling traffic capacities, queues and delays at major/minor junctions.* TRRL Report RR36. Transport and Road Research Laboratory, Crowthorne, UK.

Webster, F.V. (1958) *Traffic signal settings.* Road Research Technical Paper No. 39. The Stationery Office, London.

Chapter 6

Geometric Alignment and Design

6.1 Basic physical elements of a highway

The basic features of a highway are the carriageway itself, expressed in terms of the number of lanes used, the central reservation or median strip and the shoulders (including verges). Depending on the level of the highway relative to the surrounding terrain, side-slopes may also be a design issue.

Main carriageway

The chosen carriageway depends on a number of factors, most notably the volume of traffic using the highway, the quality of service expected from the installation and the selected design speed. In most situations a lane width of 3.65 m is used, making a standard divided or undivided 2-lane carriageway 7.3 m wide in total.

Table 6.1 gives a summary of carriageway widths normally used in the UK. These widths are as stated in TD 27/96 (DoT, 1996). Any reduction or increase in these widths is considered a departure from standard. The stated lane widths should only be departed from in exceptional circumstances such as where cyclists need to be accommodated or where the number of lanes needs to be maximised for the amount of land available. In Scotland and Northern Ireland, a total carriageway width of 6.0 m may be used on single carriageway all-purpose roads where daily flow in the design year is estimated not to exceed 5000 vehicles.

Central reservation

A median strip or central reservation divides all motorways/dual carriageways. Its main function is to make driving safer for the motorist by limiting locations where vehicles can turn right (on dual carriageways), completely separating the traffic travelling in opposing directions and providing a space where vehicles can recover their position if for some reason they have unintentionally left the carriageway. In urban settings, a width of 4.5 m is recommended for 2/3-lane dual carriageways, with 4.0 m recommended for rural highways of this type. While these values should be the first option, a need to minimise land take might

Road description	Carriageway width (m)
Urban/rural 4-lane dual	14.60
Urban/rural 3-lane dual	11.00
Urban/rural single/dual 2-lane (normal)	7.30
Rural single 2-lane (wide)	10.00

Table 6.1 Standard carriageway widths

lead to reductions in their value. Use of dimensions less than those recommended is taken as a relaxation rather than a departure from the standard (TD 27/96). (The term 'relaxation' refers to a relaxing of the design standard to a lower level design step, while a 'departure' constitutes non-adherence to a design standard where it is not realistically achievable. Both these terms are defined in more detail in section 6.3.) Use of central reservation widths greater than the values stated is permitted. Its surfacing material should be different to that on the carriageway itself. Grass, concrete or bituminous material can be used.

Hard strips/verges

On single carriageway roads (normal and wide), a 1 m wide hardstrip and a 2.5 m wide grassed verge is employed on the section of roadway immediately adjacent to the main carriageway on each side. On rural 2 and 3-lane motorways, a hardshoulder of 3.3 m and a verge of 1.5 m are the recommended standard. On rural 2/3-lane dual carriageways, the 1 m wide hardstrip and 2.5 m wide verge is detailed on the nearside with a 1 m hardstrip on the offside. For urban motorways the verge dimension varies while the hard shoulder is set at 2.75 m wide.

Diagrams of typical cross-sections for different road classifications are given in Figs 6.1 to 6.4.

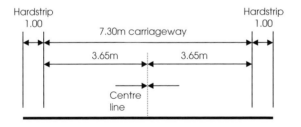

Figure 6.1 Single 7.30 metre all-purpose roadway (S2).

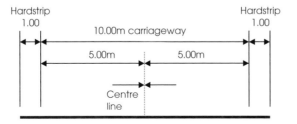

Figure 6.2 Wide single all-purpose (WS2).

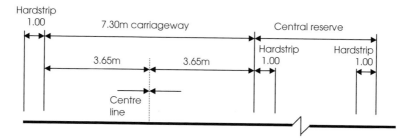

Figure 6.3 Dual 2-lane all-purpose (D2AP).

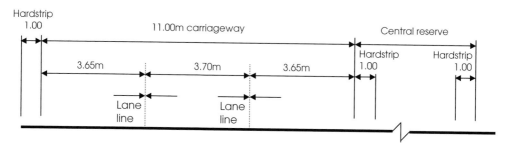

Figure 6.4 Dual 3-lane all-purpose (D3AP).

The proper geometric design of a highway ensures that drivers use the facility with safety and comfort. The process achieves this by selecting appropriate vertical and horizontal curvature along with physical features of the road such as sight distances and superelevation. The ultimate aim of the procedure is a highway that is both justifiable in economic terms and appropriate to the local environment.

6.2 Design speed, stopping and overtaking sight distances

6.2.1 Introduction

The concept of 'design speed' lies at the centre of this process. The design speed of a highway serves as a guide in the selection of the physical features referred to above. Selection of the correct design speed ensures that issues of both safety and economy in the design process are addressed. The chosen design speed must be consistent with the anticipated vehicle speeds on the highway under consideration.

The standard design speeds are 50 km/hr, 60 km/hr, 70 km/hr, 85 km/hr, 100 km/hr and 120 km/hr. These bands are based on the premise that it is considered acceptable if 85% of drivers travel at or below the designated design speed for a given highway, generally inducing a situation where approximately 99% of the drivers travel at or below one speed category above the design speed (i.e. if the speed limit is set at 85 km/hr, it can be assumed that 85% of the drivers will travel at or below this value while 99% will travel at or below 120 km/hr).

The speed bands are thus related to each other by a factor equal to the fourth root of 2, taken as being approximately 1.19. Thus, if the chosen design speed is by definition the 85th percentile speed for that highway, then the next speed band up will constitute its 99th percentile speed. The same factor separates the chosen design speed and the next speed band down, which constitutes the 50th percentile or mean speed, thus:

$$99\text{th percentile} \div 85\text{th percentile} = \sqrt[4]{2} = 1.19$$
$$85\text{th percentile} \div 50\text{th percentile} = \sqrt[4]{2} = 1.19$$

The design bands can thus be structured as shown in Table 6.2.

85th percentile speed	99th percentile speed
120	145
100	120
85	100
70	85
60	70

Table 6.2 Framework for design speeds

The geometric properties associated with the design speed of a highway constitute 'desirable values' at which 85% of the drivers are travelling with complete safety. The geometric values of the next design speed up constitute a standard at which 99% of the drivers can travel safely at the original design speed. Conversely, the geometric values of the next design speed down will constitute a relaxation at which only 50% of the drivers will be in a position to travel with complete safety at the original design speed. Such values constitute absolute minimum values. However, these may have to be adopted in difficult design conditions where many constraints both physical and otherwise exist.

Thus, in conclusion, the 85th percentile speed is selected as the design speed on the basis that it constitutes the most appropriate and judicious choice, as use of the 99th percentile would prove extremely expensive while extensive use of the 50th percentile may prove unduly unsafe for the faster travelling vehicles.

6.2.2   Urban roads

Within the UK the design speed for an urban highway is chosen on the basis of its speed limit. The value chosen will allow a small margin for speeds greater than the posted speed limit (DoT, 1993). For speed limits of 48, 64, 80 and 96 km/hr, design speeds of 60B, 70A, 85A and 100A respectively are employed.

(The suffixes A and B indicate the higher and lower categories respectively within each speed band.) These values are given in tabular form in Table 6.3. The minimum design speed for a primary distributor is set at 70A km/hr.

Table 6.3 Design speeds for urban roads

Speed limit		Design speed
mph	km/h	km/h
30	48	60B
40	64	70A
50	80	85A
60	96	100A

6.2.3   Rural roads

The design speed is determined on the basis of three factors:

- The mandatory constraint
- The layout constraint (L_c)
- The alignment constraint (A_c).

Statutory constraint

The general speed limit for motorways and dual carriageways is set at 70 mph (112 km/hr), reducing to 60 mph (96 km/hr) for single carriageways. The use of these mandatory speed limits can restrict design speeds below those freely achievable and can act as an additional constraint on speed to that dictated by the layout constraint L_c.

Layout constraint

Layout constraint assesses the degree of constraint resulting from the road cross-section, verge width and frequency of junctions and accesses. Both carriageway width and verge width are measured in metres. Density of access is expressed in terms of the total number of junctions, laybys and commercial accesses per kilometre, summed for both sides of the road using the three gradings low, medium and high, defined as:

- Low   = Between 2 and 5 accesses per kilometre
- Medium = Between 6 and 8 accesses per kilometre
- High   = Between 9 and 12 accesses per kilometre.

The layout constraints for different combinations of the above relevant parameters are defined in Table 6.4 for seven different road types.

Having sketched a trial alignment on paper, Table 6.4 is utilised to estimate L_c, whose value will range from zero for a 3-lane motorway (D3M) to 33 for a

Table 6.4 Layout constraint values (L_c)

Road type	S2				WS2		D2AP		D3AP	D2M	D3M
Carriageway width	6 m		7.3 m		10 m		Dual 7.3 m		Dual 11 m	Dual 7.3 m*	Dual 11 m*
Degree of access and junctions	H	M	M	L	M	L	M	L	L	L	L
Standard verge width	29	26	23	21	19	17	10	9	6	4	0
1.5 m verge	31	28	25	23							
0.5 m verge	33	30									

*Plus hard shoulder

6 m single carriageway road with a high level of access to it and narrow verges (S2, 6 m, H, 0.5 m verge). Where the exact conditions as defined on the table do not apply, interpolation between the given figures can be employed.

Alignment constraint

Alignment constraint measures the degree of constraint resulting from the alignment of the highway. It is assessed for both dual carriageways and single carriageways:
Dual carriageways:

$$A_c = 6.6 + B/10 \tag{6.1}$$

Single carriageways:

$$A_c = 12 - VISI/60 + 2B/45 \tag{6.2}$$

where
B   = Bendiness in degrees per kilometre (°/km)
VISI = Harmonic mean visibility

VISI can be estimated from the empirical formula:

$$Log_{10} VISI = 2.46 + VW/25 - B/400 \tag{6.3}$$

where
VW = Average verge width averaged for both sides of the road
B   = Bendiness in degrees per kilometre (°/km)

An illustration of the method for calculating bendiness in given in Fig. 6.5.
   Having determined values for the two parameters, the design speed is then estimated using Fig. 1 from TD 9/93, represented here in Fig. 6.6.

New/upgraded rural roads

In these instances, the design speed is derived in an iterative manner, with an initial alignment to a trial design speed drawn and the alignment constraint

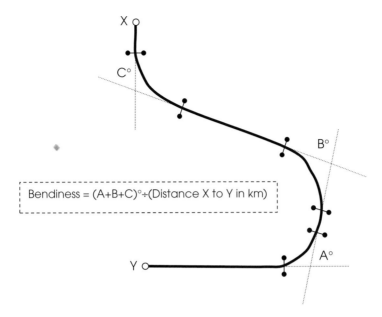

Figure 6.5 Estimation of bendiness.

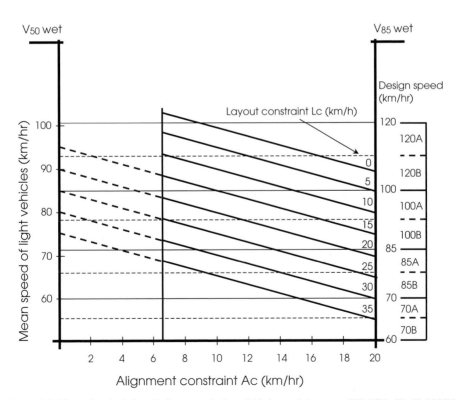

Figure 6.6 Chart for deriving design speeds (rural highways) (source: TD 9/93 (DoT, 1993)).

measured for each section of the highway over a minimum distance of 2 km. The design speed calculated from the resulting alignment and layout constraints is then checked against the originally assumed design speed so the locations can be identified where elements of the initially assumed alignment can be relaxed in order to achieve savings in terms of either cost or the environment. This procedure allows a design speed to be finalised for each section of highway under consideration. Equally, it may be necessary to upgrade the design if the resulting design speed dictates this. If any alterations to the geometry of the highway are undertaken, it will be necessary to recalculate the design speed in order to make sure that its value has not changed. The aim of this process is to ensure that all sections of the highway are both geometrically consistent and cost effective. While the design speeds for two sections running into each other need not be the same, it is advisable that their design speeds differ by no more than 10 km/hr.

Example 6.1 – Design speed calculation for an existing single carriageway route
An existing 7.3 m wide single carriageway road with 1.5 m wide verges (see Fig. 6.7) has a layout as indicated in Fig. 6.5. The length of the section of highway under examination is 4 km (X to Y). The relevant angles are:

A = 95°
B = 65°
C = 60°

There are a total of 24 access points and minor junctions along the length of the highway.
  Calculate the design speed.

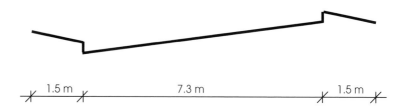

Figure 6.7 Cross-section of highway.

1.5 m   7.3 m   1.5 m

Solution

There are 24 access points along the 4 km length of roadway, therefore they occur at a rate of 6 per kilometre, giving a grading of medium.

Contd

Example 6.1 Contd

Given a 7.3 m wide single carriageway road with a verge width of 1.5 m, Table 6.4 gives a layout constraint value of 25.

$L_c = 25$

Bendiness = (95 + 65 + 60)/4
$$= 235/4$$
$$= 55°$$

The verge width (VW) is set at 1.5 m.

Therefore, the harmonic visibility (VISI) is calculated using Equation 6.3:

Log_{10} VISI = 2.46 + VW/25 − B/400
$$= 2.46 + 1.5/25 − 55/400$$
$$= 2.46 + 0.06 − 0.1375$$
$$= 2.3825$$
$$VISI = 241.27$$

The alignment constraint is then calculated using Equation 6.2:

$A_c = 12 − $ VISI/60 + 2B/45
$$= 12 − 241.27/60 + (2 × 55)/45$$
$$= 10.4$$

From Fig. 6.6, a design speed of 100 km/hr is selected (the mandatory speed limit for this class of highway is 96 km/hr).

Example 6.2 – Comparison of observed speeds with calculated design speed
Taking the existing stretch of highway referred to in Example 6.1, Table 6.5 shows the results from a speed survey taken along the route.

Table 6.5 Speed survey

Speed range (km/hr)	Observed cars
Less than 60	15
60–64	10
65–69	16
70–74	101
75–79	140
80–84	196
85–89	62
90–94	15
95–99	6
Greater than 100	1

Contd

Example 6.2 Contd

Determine the 85th percentile speed and compare it with the derived design speed.

Solution

Speed range (km/hr)	Observed cars with speed within or below this range	Percentile speed
Less than 60	15	3rd
60–64	25	4th
65–69	41	7th
70–74	142	25th
75–79	282	50th
80–84	478	85th
85–89	540	96th
90–94	555	98th
95–99	561	99th
Greater than 100	562	100th

Table 6.6 Percentiles for observed speed ranges

From the figures in Table 6.6 it can be seen that the 85th percentile speed is in the range of 80–84 km/hr. Thus the observed driver speeds are appreciably below the design speed/mandatory speed limit that has been allowed for and, as a consequence, the road would not be a priority for upgrading/improvement.

6.3   Geometric parameters dependent on design speed

For given design speeds, designers aim to achieve at least the desirable minimum values for stopping sight distance, horizontal curvature and vertical crest curves. However, there are circumstances where the strict application of desirable minima would lead to disproportionately high construction costs or environmental impact. In such situations either of two lower tiers can be employed:

- Relaxations
- Departures.

Relaxations

This second tier of values will produce a level of service that may remain acceptable and will lead to a situation where a highway may not become unsafe. The limit for relaxations is defined by a set number of design speed steps below a benchmark level – usually the desirable minimum (TD 9/93). Relaxations can be used at the discretion of the designer, having taken into consideration appropriate local factors and advice in relevant documentation.

Departures

In situations of exceptional difficulty where even a move to the second tier in the hierarchy, i.e. relaxations, cannot resolve the situation, adoption of a value within the third tier of the hierarchy – a departure – may have to be considered. Care must be taken that safety is not significantly reduced. In order for a departure from standard to be adopted for a major road scheme, the designer must receive formal approval from central government or its responsible agency before it can be incorporated into the design layout.

6.4   Sight distances

6.4.1   Introduction

Sight distance is defined as the length of carriageway that the driver can see in both the horizontal and vertical planes. Two types of sight distance are detailed: stopping distance and overtaking distance.

6.4.2   Stopping sight distance

This is defined as the minimum sight distance required by the driver in order to be able to stop the car before it hits an object on the highway. It is of primary importance to the safe working of a highway.

Table 6.7 indicates the stopping sight distances for the different design speeds. Both desirable minimum and absolute minimum values are given. As seen, the latter category constitutes in each case a relaxation equal to one design speed step below the desirable minimum.

The standard TD 9/93 requires stopping sight distance to be measured from a driver's eye height of between 1.05 m and 2 m above the surface of the highway to an object height of between 0.26 m and 2 m above it. These values ensure that drivers of low-level cars can see small objects on the carriageway ahead. The

Table 6.7 Stopping sight distances for different design speeds (source: TD 9/93 (DoT, 1993))

Stopping sight distance (m)	Design speed (km/hr)					
	120	100	85	70	60	50
Desirable minimum	295	215	160	120	90	70
Absolute minimum	215	160	120	90	70	50

vast majority (>95%) of driver heights will be greater than 1.05 m while, at the upper range, 2 m is set as the typical eye height for the driver of a large heavy goods vehicle. With regard to object heights, the range 0.26 m to 2 m is taken as encompassing all potential hazards on the road.

Checks should be carried out in both the horizontal and vertical planes. This required envelope of visibility is shown in Fig. 6.8.

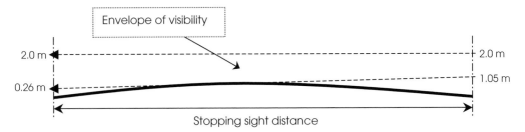

Figure 6.8 Measurement of stopping sight distance.

The distance itself can be subdivided into three constituent parts:

- The perception distance – length of highway travelled while driver perceives hazard
- The reaction distance – length of highway travelled during the period of time taken by the driver to apply the brakes and for the brakes to function
- The braking distance – length of highway travelled while the vehicle actually comes to a halt.

The combined perception and reaction time, t, can vary widely depending on the driver. However, in the UK, a value of 2 seconds is taken as being appropriate for safe and comfortable design.

The length of highway travelled during the perception-reaction time is calculated from the formula:

$$\text{Perception-reaction distance (m)} = 0.278tV \tag{6.4}$$

where
V = initial speed (km/hr)
t = combined perception and reaction time (s)

A rate of deceleration of $0.25g$ is generally used for highway design in the UK. This value can be achieved on normally-textured surfaces in wet conditions without causing discomfort to the driver and passengers.

$$\text{Braking distance (m)} = v^2/2w \tag{6.5}$$

where
v = initial speed (m/s)
w = rate of deceleration (m/s^2)

Combining Equations 6.4 and 6.5 with the appropriate design speeds and rounding the resulting values as appropriate yields the set of stopping sight distances given in Table 6.7.

6.4.3   Overtaking sight distance

Overtaking sight distance is of central importance to the efficient working of a given section of highway. Overtaking sight distance only applies to single carriageways. There is no full overtaking sight distance (FOSD) for a highway with a design speed of 120 km/hr since this design speed is not suitable for a single carriageway road.

Full overtaking sight distances are much larger in value than stopping sight distances. Therefore, economic realities dictate that they can only be complied with in relatively flat terrain where alignments, both vertical and horizontal, allow the design of a relatively straight and level highway.

Values for different design speeds are given in Table 6.8.

Table 6.8 Full overtaking sight distances for different design speeds (source: TD 9/93 (DOT, 1993))

| | Design speed (km/hr) | | | | | |
	120	100	85	70	60	50
Full overtaking sight distance (m)	—	580	490	410	345	290

Full overtaking sight distance is measured from vehicle to vehicle (the hazard or object in this case is another car) between points 1.05 m and 2.00 m above the centre of the carriageway. The resulting envelope of visibility for this set of circumstances is shown in Fig. 6.9.

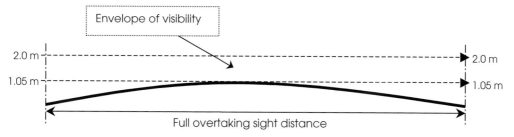

Figure 6.9 Measurement of full overtaking sight distance (FOSD).

Full overtaking sight distance is made up of three components: d_1, d_2 and d_3.

d_1 = Distance travelled by the vehicle in question while driver in the overtaking vehicle completes the passing manoeuvre (Overtaking Time)

d_2 = Distance between the overtaking and opposing vehicles at the point in time at which the overtaking vehicle returns to its designated lane (Safety Time)

d_3 = Distance travelled by the opposing vehicle within the above mentioned 'perception-reaction' and 'overtaking' times (Closing Time).

In order to establish the values for full overtaking sight distance, it is assumed that the driver making the overtaking manoeuvre commences it at two design speed steps below the designated design speed of the section of highway in question. The overtaking vehicle then accelerates to the designated design speed. During this time frame, the approaching vehicle is assumed to travel towards the overtaking vehicle at the designated design speed. d_2 is assumed to be 20% of d_3.

These assumptions yield the following equation:

$$FOSD = 2.05tV \qquad (6.6)$$

where

V = design speed (m/s)

t = time taken to complete the entire overtaking manoeuvre (s)

The value of t is generally taken as 10 seconds, as it has been established that it is less than this figure in 85% of observed cases.

If we are required to establish the FOSD for the 85th percentile driver on a section of highway with a design speed of 85 km/hr (23.6 m/s), we can use Equation 6.6 as follows:

$$FOSD^{85} = 2.05 \times 10 \times 23.6$$
$$= 483.8 \, m$$

This figure is a very small percentage less than the value given in TD 9/93 and illustrated in Table 6.8 (490 m).

If we go back to the three basic components of FOSD, d_1, d_2, and d_3, we can derive a very similar value:

d_1 = 10 seconds travelling at an average speed of 70 km/hr (19.4 m/s)
    = 10×19.4 m
    = 194 m

d_3 = Opposing vehicle travels 10×23.6 m
    = 236 m

d_2 = $d_3/5$
    = 47.2 m

FOSD = 194 + 236 + 47.2
    = 477.2 m, which is within approximately 1% of the value derived from Equation 6.6.

It is imperative that, in the interests of safety, along a given stretch of highway there is no confusion on the driver's part as to whether or not it is safe to overtake. On stretches where overtaking is allowed, the minimum values given in

Table 6.8 should be adhered to. Where overtaking is not permitted, sight distances should not greatly exceed those required for safe stopping.

6.5 Horizontal alignment

6.5.1 General

Horizontal alignment deals with the design of the directional transition of the highway in a horizontal plane. A horizontal alignment consists, in its most basic form, of a horizontal arc and two transition curves forming a curve which joins two straights. In certain situations the transition curve may have zero length. The design procedure itself must commence with fixing the position of the two straight lines which the curve will join together. The basic parameter relating these two lines is the intersection angle. Figure 6.10 indicates a typical horizontal alignment.

Minimum permitted horizontal radii depend on the design speed and the superelevation of the carriageway, which has a maximum allowable value of 7% in the UK, with designs in most cases using a value of 5%. The relationship between superelevation, design speed and horizontal curvature is detailed in the subsection below.

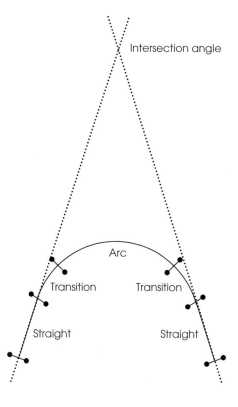

Figure 6.10 Typical horizontal alignment.

Table 6.9 Horizontal radii for different design speeds and superelevation, *e*
(source: TD 9/93 (DOT, 1993))

Horizontal curvature (R)	Design speed (km/hr)					
	120	100	85	70	60	50
Minimum *R* with *e* = 2.5% (not recommended for single carriageways)	2040	1440	1020	720	510	360
Minimum *R* with *e* = 3.5% (not recommended for single carriageways)	1440	1020	720	510	360	255
Desirable minimum *R* with *e* = 5% (m)	1020	720	510	360	255	180
Absolute minimum *R* with *e* = 7% (m)	720	510	360	255	180	127
One step below absolute minimum with *e* = 7%	510	360	255	180	127	90

Table 6.9 details the minimum radii permitted for a given design speed and value of superelevation which should not exceed 7%.

6.5.2 Deriving the minimum radius equation

Figure 6.11 illustrates the forces acting on a vehicle of weight W as it is driven round a highway bend of radius R. The angle of incline of the road (superelevation) is termed α. P denotes the side frictional force between the vehicle and the highway, and N the reaction to the weight of the vehicle normal to the surface of the highway. C is the centrifugal force acting horizontally on the vehicle and equals $M \times v^2/R$ where M is the mass of the vehicle.

As all the forces in Figure 6.11 are in equilibrium, they can be resolved along the angle of inclination of the road:

(Weight of vehicle resolved parallel to highway) + (Side friction factor)
= (Centrifugal force resolved parallel to highway)

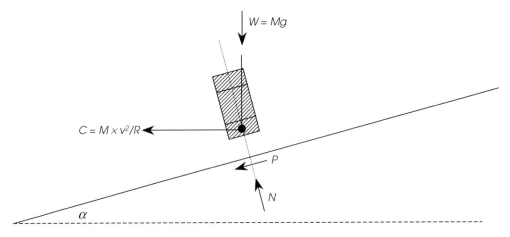

Figure 6.11 Forces on a vehicle negotiating a horizontal curve.

$$[Mg \times \text{Sin}(\alpha)] + P = [(M \times v^2/R) \times \text{Cos}(\alpha)] \qquad (6.7)$$

The side frictional force, P, can be expressed as:

$$P = \mu[W \times \text{Cos}(\alpha) + C \times \text{Sin}(\alpha)]$$
$$= \mu[Mg \times \text{Cos}(\alpha) + M \times v^2/R \times \text{Sin}(\alpha)] \qquad (6.8)$$
(μ is defined as the side friction factor)

Substituting Equation 6.8 into Equation 6.7, the following expression is derived:

$$[Mg \times \text{Sin}(\alpha)] + \mu[Mg \times \text{Cos}(\alpha) + M \times v^2/R \times \text{Sin}(\alpha)] = [(M \times v^2/R) \times \text{Cos}(\alpha)]$$

Dividing across by $Mg\ \text{Cos}(\alpha)$:

$$\tan(\alpha) + \mu + \mu\, v^2/gR \tan(\alpha) = v^2/gR \qquad (6.9)$$

If we ignore the term $\mu v^2/gR\tan(\alpha)$ on the basis that it is extremely small, the following final expression is derived:

$$\tan(\alpha) + \mu = v^2/gR$$

The term $\tan(\alpha)$ is in fact the superelevation e. If in addition we express velocity in kilometres per hour rather than metres per second, and given that g equals $9.81\,\text{m/s}^2$, the following generally used equation is obtained:

$$\frac{V^2}{127R} = e + \mu \qquad (6.10)$$

This expression is termed the minimum radius equation. It is the formula which forms the basis for the values of R illustrated in Table 6.9.

In UK design practice, it is assumed that, at the design speed, 55% of the centrifugal force is balanced by friction, with the remaining 45% being counteracted by the crossfall. Thus, Equation 6.10 becomes:

$$e = \frac{0.45 \times V^2}{127R} \quad \text{or}$$
$$e = \frac{0.353V^2}{R} \ (e \text{ is expressed in percentage terms}) \qquad (6.11)$$

Therefore:

$$R = \frac{0.353V^2}{e} \qquad (6.12)$$

Therefore, assuming e has a value of 5% (appropriate for the desirable minimum radius R):

$$R = 0.07069V^2 \qquad (6.13)$$

Taking a design speed of 120 km/hr:

$$R = 0.07069(120)^2$$
$$= 1018\,\text{m}$$

(The appropriate value for desirable minimum radius given by TD 9/93 and illustrated in Table 6.9 is almost identical – 1020 m.)

Taking a design speed of 85 km/hr:

$$R = 0.07069(85)^2$$
$$= 510.7 \text{ m}$$

(Again, the appropriate value for desirable minimum radius given by TD 9/93 and illustrated in Table 6.9 is almost identical – 510 m.)

It can be seen that, as with sight distances, the absolute minimum values of R, consistent with a superelevation of 7%, are one design step below the desirable minimum levels. These are termed *relaxations*. One further design step below these absolute minima, also termed limiting radius values, is given in Table 6.9. These represent *departures* and are associated with situations of exceptional difficulty.

6.5.3   *Horizontal curves and sight distances*

It is imperative that adequate sight distance be provided when designing the horizontal curves within a highway layout. Restrictions in sight distance occur when obstructions exist, as shown in Fig. 6.12. These could be boundary walls or, in the case of a section of highway constructed in cut, an earthen embankment.

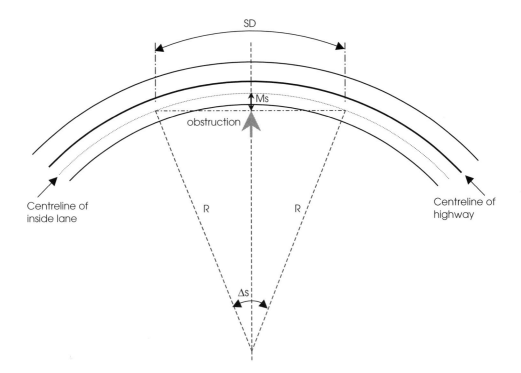

Figure 6.12 Required clearance for sight distance on horizontal curves.

The minimum offset clearance Ms required between the centreline of the highway and the obstruction in question can be estimated in terms of the required sight distance SD and the radius of curvature of the vehicle's path R as follows.

It is assumed that the sight distance lies within the length of the horizontal curve. The degree of curve is defined as the angle subtended by a 100 m long arc along the horizontal curve. It measures the sharpness of the curve and can be related to the radius of the curve as follows:

$$D = 5729.6 \div R \tag{6.14}$$

An analysis of the geometry yields the following formula relating the length of a curve to the degree of curve:

$$L = 100 \times \Delta s \div D \tag{6.15}$$

where Δ is the central angle of the curve.

Assuming in this case that the length of the curve is SD, Equation 6.15 can be written:

$$SD = 100 \times \Delta s \div D \tag{6.16}$$

Substituting Equation 6.14 into the above equation yields:

$$\Delta s = 57.296 \times SD \div R \tag{6.17}$$

Since:

$$Cos\left(\frac{\Delta s}{2}\right) = (R - Ms) \div R$$

Therefore:

$$Ms = R[1 - Cos(\Delta s / 2)] \tag{6.18}$$

Substituting Equation 6.17 into 6.18 the following equation is obtained:

$$Ms = R[1 - Cos(28.65 \times SD/R)] \tag{6.19}$$

Example 6.3

A 2-lane 7.3 m wide single carriageway road has a curve radius of 600 m. The minimum sight stopping distance required is 160 m.

Calculate the required distance to be kept clear of obstructions in metres.

Solution

Applying Equation 6.19:

$$Ms = 600[1 - Cos(28.65 \times 160/600)]$$
$$= 5.33 \, m$$

Alternative method for computing Ms

If the radius of horizontal curvature is large, then it can be assumed that SD approximates to a straight line. Therefore, again assuming that the sight distance length SD lies within the curve length, the relationship between R, Ms and SD can be illustrated graphically as shown in Fig. 6.13.

Using the right-angle rule for triangle A in Fig. 6.13:

$$R^2 = x^2 + (R - Ms)^2$$

Therefore:

$$x^2 = R^2 - (R - Ms)^2 \tag{6.20}$$

Now, again using the right-angle rule, this time for triangle B in Fig. 6.13:

$$(SD/2)^2 = x^2 + Ms^2$$

Therefore:

$$x^2 = (SD/2)^2 - Ms^2 \tag{6.21}$$

Combining Equations 6.20 and 6.21:

$$(SD/2)^2 - Ms^2 = R^2 - (R - Ms)^2$$

Therefore:

$$(SD/2)^2 - Ms^2 = R^2 - (R^2 + Ms^2 - 2 \times R \times Ms)$$

Cancelling out the R^2 and Ms^2 terms:

$$(SD/2)^2 = 2 \times R \times Ms$$

Therefore:

$$Ms = SD^2/8R \tag{6.22}$$

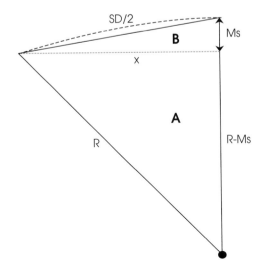

Figure 6.13
Horizontal curve/SD relationship (assuming SD to be measured along straight).

Example 6.4

Using the same data as in Example 6.3, calculate the value of Ms using the approximate method.

Solution

Applying Equation 6.22:

Ms = 160²/(8 × 600)
     = 25600/4800
     = 5.33 m

If the sight distance length SD lies outside the curve length R, the following formula can be derived for estimating the minimum offset clearance:

$$Ms = L(2 \times SD - L)/8R \qquad (6.23)$$

For a derivation of this equation, see O'Flaherty (1986).

6.5.4   Transitions

These curve types are used to connect curved and straight sections of highway. (They can also be used to ease the change between two circular curves where the difference in radius is large.) The purpose of transition curves is to permit the gradual introduction of centrifugal forces. Such forces are required in order to cause a vehicle to move round a circular arc rather than continue in a straight line. A finite quantity of time, long enough for the purposes of ease and safety, will be required by the driver to turn the steering wheel. The vehicle will follow its own transition curve as the driver turns the steering wheel. The radial acceleration experienced by the vehicle travelling at a given velocity v changes from zero on the tangent to v^2/R when on the circular arc. The form of the transition curve should be such that the radial acceleration is constant.

The radius of curvature of a transition curve gradually decreases from infinity at the intersection of the tangent and the transition curve to the designated radius R at the intersection of the transition curve with the circular curve.

Transition curves are normally of spiral or clothoid form:

$$RL = A^2$$

where
A^2 is a constant that controls the scale of the clothoid
R is the radius of the horizontal curve
L is the length of the clothoid

Two formulae are required for the analysis of transition curves:

$$S = L^2/24R \tag{6.24}$$

$$L = V^3/(3.6^3 \times C \times R) \tag{6.25}$$

where
S is the shift (m)
L is the length of the transition curve (m)
R is the radius of the circular curve (m)
V is the design speed (km/hr)
C is the rate of change of radial acceleration (m/s³)

The value of C should be within the range 0.3 to 0.6. A value above 0.6 can result in instability in the vehicle while values less than 0.3 will lead to excessively long transition curves leading to general geometric difficulties. The design process usually commences with an initial value of 0.3 being utilised, with this value being increased gradually if necessary towards its upper ceiling.

The length of transition should normally be limited to $(24R)^{0.5}$ (TD 9/93), thus:

$$L_{\mathrm{max}} = \sqrt{24R} \tag{6.26}$$

Shift

Figure 6.14 illustrates the situation where transition curves are introduced between the tangents and a circular curve of radius R. Here, the circular curve

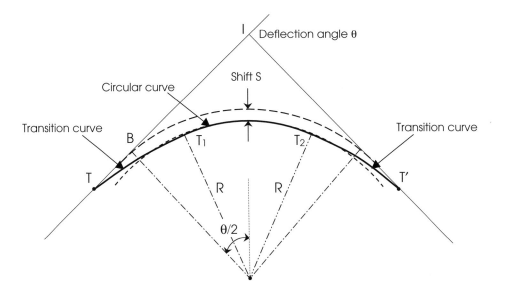

Figure 6.14 Transition curves.

must be shifted inwards from its initial position by the value S so that the curves can meet tangentially. This is the same as having a circular curve of radius ($R + S$) joining the tangents replaced by a circular curve (radius R) and two transition curves. The tangent points are, however, not the same. In the case of the circular curve of radius ($R + S$), the tangent occurs at B, while for the circular/transition curves, it occurs at T (see Fig. 6.14).

From the geometry of the above figure:

$$IB = (R + S)\tan(\theta/2) \tag{6.27}$$

It has been proved that B is the mid-point of the transition (see Bannister and Raymond, 1984 for details).

Therefore:

$$BT = L/2 \tag{6.28}$$

Combining these two equations, the length of the line IT is obtained:

$$IT = (R + S)\tan(\theta/2) + L/2 \tag{6.29}$$

If a series of angles and chord lengths are used, the spiral is the preferred form. If, as is the case here, x and y co-ordinates are being used, then any point on the transition curve can be estimated using the following equation of the curve which takes the form of a cubic parabola (see Fig. 6.15):

$$x = y^3 \div 6RL \tag{6.30}$$

When y attains its maximum value of L (the length of the transition curve), then the maximum offset is calculated as follows:

$$x = L^3 \div 6RL = L^2 \div 6R \tag{6.31}$$

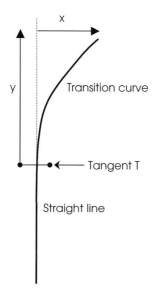

Figure 6.15
Generation of offset values for plotting a transition curve.

Example 6.5

A transition curve is required for a single carriageway road with a design speed of 85 km/hr. The bearings of the two straights in question are 17° and 59° (see Fig. 6.16). Assume a value of 0.3 m/s³ for C.

Calculate the following:

(1) The transition length, L
(2) The shift, S
(3) The length along the tangent required from the intersection point to the start of the transition, IT
(4) The form of the cubic parabola and the co-ordinates of the point at which the transition becomes the circular arc of radius R.

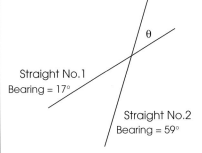

Straight No.1
Bearing = 17°

Straight No.2
Bearing = 59°

Figure 6.16
Intersection angle θ
between straights.

Solution

The design speed is 85 km/hr, therefore the desirable minimum radius is 510 m, assuming superelevation of 5%.
Length of transition:
Using Equation 6.25:

$$L = (85)^3/(3.6^3 \times 0.3 \times 510)$$
$$= 86.03 \text{ m}$$

Note: Equation 6.26 dictates that the transition be no longer than $(24R)^{0.5}$. In this case:

$$L_{\text{max}} = \sqrt{24R} = \sqrt{24 \times 510} = 110.6 \text{ m} > 86.03 \text{ m}$$

Therefore the derived length is less than the maximum permissible value.

Contd

Example 6.5 Contd

Shift:
Using Equation 6.24:

$$S = L^2/24R = (86.03)^2 \div 24 \times 510$$
$$= 0.605\,m$$

Length of IT:
Using Equation 6.29

$$IT = (R+S)\tan(\theta/2) + L/2$$
$$= (510.605)\tan(42/2) + 86.03/2$$
$$= 196.00 + 43.015$$
$$= 239.015\,m$$

Form of the transition curve:
Using Equation 6.31
$$x = y^3 \div 6RL$$
$$= y^3 \div 6 \times 510 \times 86.03$$
$$= y^3 \div 263\,251.8 \tag{6.32}$$

Co-ordinates of point at which circular arc commences:
This occurs where y equals the transition length (86.03 m).
   At this point, using Equation 6.31:

$$x = (86.03)^2 \div (6 \times 510)$$
$$= 2.419\,m$$

This point can now be fixed at both ends of the circular arc. Knowing its radius we are now in a position to plot the circle.

Note: In order to actually plot the curve, a series of offsets must be generated. The offset length used for the intermediate values of y is typically between 10 and 20 m. Assuming an offset length of 10 m, the values of x at any distance y along the straight joining the tangent point to the intersection point, with the tangent point as the origin (0,0), are as shown in Table 6.10, using Equation 6.32.

y	x
10	0.0038
20	0.0304
30	0.1026
40	0.2431
50	0.4748
60	0.8205
70	1.303
80	1.945

Table 6.10 Offsets at 10 m intervals

6.6   Vertical alignment

6.6.1   General

Once the horizontal alignment has been determined, the vertical alignment of the section of highway in question can be addressed. Again, the vertical alignment is composed of a series of straight-line gradients connected by curves, normally parabolic in form (see Fig. 6.17). These vertical parabolic curves must therefore be provided at all changes in gradient. The curvature will be determined by the design speed, being sufficient to provide adequate driver comfort with appropriate stopping sight distances provided.

The desirable maximum vertical gradients are shown in Table 6.11.

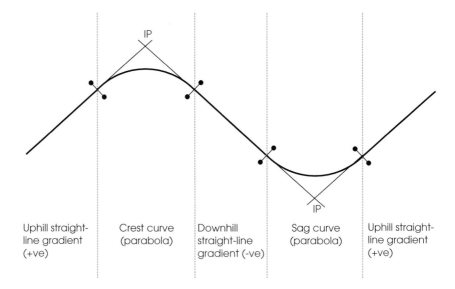

Figure 6.17 Example of typical vertical alignment.

Road type	Desirable maximum gradient (%)
Motorway	3
All-purpose dual carriageway	4
All-purpose single carriageway	6

Table 6.11 Desirable maximum vertical gradients

In difficult terrain, use of gradients steeper than those given in Table 6.11 may result in significant construction and/or environmental savings. The absolute maximum for motorways is 4%. This threshold rises to 8% for all-purpose roads, with any value above this considered a departure from standards (DoT, 1993). A minimum longitudinal gradient of 0.5% should be maintained where possible in order to ensure adequate surface water drainage. (This can also be dealt with through the provision of a drainage system running parallel to the highway.)

6.6.2   K values

The required minimum length of a vertical curve is given by the equation:

$$L = K(p - q) \tag{6.33}$$

K is a constant related to design speed. K values are given in Table 6.12.

Table 6.12 K values for vertical curvature

	Design speed (km/hr)					
	120	100	85	70	60	50
Desirable minimum K value – Crest curves (not recommended for single carriageways)	182	100	55	30	17	10
Absolute minimum K value – Crest curves	100	55	30	17	10	6.5
Absolute minimum K value – Sag curves	37	26	20	20	13	9
Full overtaking sight distance (FOSD) K value – Crest curve	—	400	285	200	142	100

Example 6.6

Calculate the desired and absolute minimum crest curve lengths for a dual carriageway highway with a design speed of 100 km/hr where the algebraic change in gradient is 7% (from +3% (uphill) to −4% (downhill)).

Solution

From Table 6.12, the appropriate K values are 100 and 55.

(1)   Desirable minimum curve length = 100 × 7 = 700 m
(2)   Absolute minimum curve length = 55 × 7 = 385 m

6.6.3   Visibility and comfort criteria

Desirable minimum curve lengths in this instance are based on visibility concerns rather than comfort as, above a design speed of 50 km/hr, the crest in the road will restrict forward visibility to the desirable minimum stopping sight distance before minimum comfort criteria are applied (TD 9/93). With sag curves, as visibility is, in most cases, unobstructed, comfort criteria will apply. Sag curves should therefore normally be designed to the absolute minimum K value detailed in Table 6.12. For both crest and sag curves, relaxations below the desired minimum values may be made at the discretion of the designer, though the number of design steps permitted below the desirable minimum value will vary depending on the curve and road type, as shown in Table 6.13.

Road type	Crest curve	Sag curve
Motorway	1 or 2 steps	0 steps
All-purpose	2 or 3 steps	1 or 2 steps

Table 6.13 Permitted relaxations for different road and vertical curve types (below desired min. for crest curves and below absolute min. for sag curves)

6.6.4   Parabolic formula

Referring to Fig. 6.18, the formula for determining the co-ordinates of points along a typical vertical curve is:

$$y = \left[\frac{q-p}{2L}\right]x^2 \qquad (6.34)$$

where
p and q are the gradients of the two straights being joined by the vertical curve in question.
L is the vertical curve length
x and y are the relevant co-ordinates in space

Proof

If Y is taken as the elevation of the curve at a point x along the parabola, then:

$$\frac{d^2Y}{dx^2} = k(a\ constant) \qquad (6.35)$$

Integrating Equation 6.35:

$$\frac{dY}{dx} = kx + C \qquad (6.36)$$

Examining the boundary conditions:
When x = 0:

$$\frac{dY}{dx} = p \qquad (6.37)$$

(p being the slope of the first straight line gradient)
Therefore:

$$p = C \qquad (6.38)$$

When x = L:

$$\frac{dY}{dx} = q \tag{6.39}$$

(q being the slope of the second straight line gradient)
Therefore:

$$q = kL + C = kL + p \tag{6.40}$$

Rearranging Equation 6.40:

$$k = (q - p) \div L \tag{6.41}$$

Substituting Equations 6.38 and 6.41 into Equation 6.36:

$$\frac{dY}{dx} = \left(\frac{q-p}{L}\right)x + p \tag{6.42}$$

Integrating Equation 6.42:

$$Y = \left(\frac{q-p}{L}\right)\frac{x^2}{2} + px \tag{6.43}$$

From Fig. 6.18:

$$p = (y + Y) \div x \tag{6.44}$$

Substituting Equation 6.44 into Equation 6.43:

$$Y = \left(\frac{q-p}{L}\right)\frac{x^2}{2} + (y + Y) \tag{6.45}$$

Rearranging Equation 6.45:

$$y = -\left(\frac{q-p}{2L}\right)x^2$$

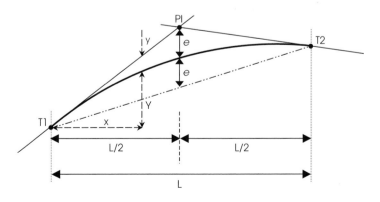

Figure 6.18 Basic parabolic curve.

where x is the distance along the curve measured from the start of the vertical curve and y is the vertical offset measured from the continuation of the slope to the curve.

At the intersection point PI:

$$x = L/2$$

Therefore

$$e = -\left(\frac{q-p}{2L}\right)\left(\frac{L}{2}\right)^2 = y$$

$$= -(q-p)\frac{L}{8} \tag{6.46}$$

The co-ordinates of the highest/lowest point on the parabolic curve, frequently required for the estimation of minimum sight distance requirements, are:

$$x = \frac{Lp}{p-q} \tag{6.47}$$

$$y = \frac{Lp^2}{2(p-q)} \tag{6.48}$$

Example 6.7

A vertical alignment for a single carriageway road consists of a parabolic crest curve connecting a straight-line uphill gradient of +4% with a straight-line downhill gradient of −3%.

(1) Calculate the vertical offset at the point of intersection of the two tangents at PI
(2) Calculate the vertical and horizontal offsets for the highest point on the curve.

Assume a design speed of 85 km/hr and use the absolute minimum K value for crest curves.

Solution

Referring to Table 6.12, a K value of 30 is obtained. This gives an absolute minimum curve length of 210 m.

Vertical offset at PI:
    p = +4%
    q = −3%

Contd

Example 6.7 Contd

Using Equation 6.46:

$$e = -(q-p)\frac{L}{8} = -(-0.03-(0.04)\times 210) \div 8$$

$$= 1.8375\,m$$

Co-ordinates of highest point on crest curve:
Using Equations 6.47 and 6.48

$$x = \frac{Lp}{p-q} = (210\times 0.04) \div (0.04+0.03)$$

$$= 120\,m$$

$$y = \frac{Lp^2}{2(p-q)} = (210\times 0.04^2) \div 2 \times (0.04+0.03)$$

$$= 2.4\,m$$

Since, from Equation 6.44

$$p = (y+Y) \div x$$
$$Y = px - y$$
$$= 0.04(120) - 2.4$$
$$= 2.4\,m$$

6.6.5   Crossfalls

To ensure adequate rainfall run-off from the surface of the highway, a minimum crossfall of 2.5% is advised, either in the form of a straight camber extending from one edge of the carriageway to the other or as one sloped from the centre of the carriageway towards both edges (see Fig. 6.19).

6.6.6   *Vertical crest curve design and sight distance requirements*

In the case of a crest curve, the intervening highway pavement obstructs the visibility between driver and object. The curvature of crest curves should be sufficiently large in order to provide adequate sight distance for the driver. In order to provide this sight distance, the curve length L is a critical parameter. Too great a length is costly to the developer while too short a length compromises critical concerns such as safety and vertical clearance to structures.

For vertical crest curves, the relevant parameters are:

- The sight distance S
- The length of the curve L

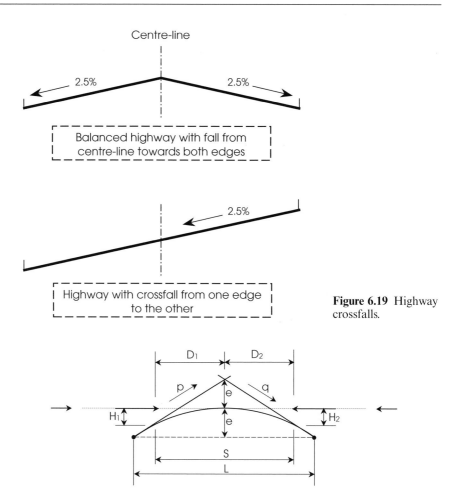

Figure 6.19 Highway crossfalls.

Figure 6.20 Case (1) $S \leq L$.

- The driver's eye height H_1
- The height of the object on the highway H_2
- Minimum curve length L_m.

In order to estimate the minimum curve length, L_m, of a crest curve, two conditions must be considered. The first, illustrated in Fig. 6.20, is where the required sight distance is contained within the crest curve length ($S \leq L$), while the second (see Fig. 6.21) entails the sight distance extending into the tangents either side of the parabolic crest curve ($S > L$).

The formulae relating to these two conditions are:

$$L_m = \frac{AS^2}{\left[\sqrt{2H_1} + \sqrt{2H_2}\right]^2} \text{ for } (S \leq L) \tag{6.49}$$

$$L_m = 2S - \frac{2\left[\sqrt{H_1} + \sqrt{H_2}\right]^2}{A} \text{ for } (S > L) \tag{6.50}$$

where
A is the algebraic difference between the two straight-line gradients.

Derivation of crest curve formulae

Case (1) $S \leq L$

Given that the curve is parabolic, the relevant offsets are equal to a constant times the square of the distance from the point at which the crest curve is tangential to the line of sight. Thus, with reference to Fig. 6.20:

$$H_1 = k(D_1)^2 \tag{6.51}$$

And:

$$H_2 = k(D_2)^2 \tag{6.52}$$

Since $e = k(L/2)^2$:

$$\frac{H_1 + H_2}{e} = \frac{4(D_1)^2 + 4(D_2)^2}{L^2} \tag{6.53}$$

Thus:

$$D_1 + D_2 = \sqrt{\frac{H_1 L^2}{4e}} + \sqrt{\frac{H_2 L^2}{4e}} \tag{6.54}$$

From Equation 6.46:

$$e = \frac{LA}{8}$$

Therefore, substituting this expression into Equation 6.54:

$$D_1 = \sqrt{\frac{2H_1 L}{A}} \tag{6.55}$$

And:

$$D_2 = \sqrt{\frac{2H_2 L}{A}} \tag{6.56}$$

Bringing L over to the RHS of the equation:

$$L = \frac{A(D_1 + D_2)^2}{(\sqrt{2H_1} + \sqrt{2H_2})^2} \tag{6.57}$$

Since S, the required sight distance, equals $D_1 + D_2$:

$$L = L_m = \frac{AS^2}{(\sqrt{2H_1} + \sqrt{2H_2})^2} \quad \text{(see Equation 6.49)}$$

If the object is assumed to have zero height ($H_2 = 0$), then Equation 6.49 is reduced to:

$$L = \frac{AS^2}{2H_1} \qquad (6.58)$$

If the object is assumed to be at the driver's eye height ($H_1 = H_2$):

$$L = \frac{AS^2}{8H_1} \qquad (6.59)$$

Case (2) S > L

Referring to Fig. 6.21, if we assume that g is equal to the difference between the slope of the sight line and the slope of the rising straight-line gradient, p, then the sight distance S can be estimated as follows:

$$S = \frac{L}{2} + \frac{H_1}{g} + \frac{H_2}{A - g} \qquad (6.60)$$

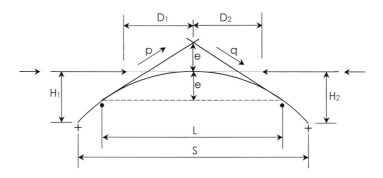

Figure 6.21 Case (2) $S > L$.

In order to derive the minimum sight distance S_{min}, S is differentiated with respect to g as follows:

$$\frac{dS}{dg} = -\frac{H_1}{g^2} + \frac{H_2}{(A - g)^2} = 0 \qquad (6.61)$$

Therefore:

$$g = \frac{A\sqrt{H_1 H_2} - H_1 A}{H_2 - H_1} \qquad (6.62)$$

Substituting Equation 6.62 into Equation 6.60:

$$S = \frac{L}{2} + \left\{ H_1 \div \left[\frac{A\sqrt{H_1 H_2} - H_1 A}{H_2 - H_1} \right] \right\} + \left\{ H_2 \div \left[A - \frac{A\sqrt{H_1 H_2} - H_1 A}{H_2 - H_1} \right] \right\} \qquad (6.63)$$

Bringing L over to the left-hand side of the equation:

$$L = L_{\mathrm{m}} = 2S - \frac{2(\sqrt{H_1} + \sqrt{H_2})^2}{A} \text{ (see Equation 6.50)}$$

If the object is assumed to have zero height ($H_2 = 0$), then Equation 6.50 is reduced to:

$$L = 2S - \frac{2H_1}{A} \tag{6.64}$$

If the object is assumed to be at the driver's eye height ($H_1 = H_2$):

$$L = 2S - \frac{8H_1}{A} \tag{6.65}$$

Example 6.8

A vertical crest curve on a single carriageway road with a design speed of 85 km/hr is to be built in order to join an ascending grade of 4% with a descending grade of 2.5%. The motorist's eye height is assumed to be 1.05 m while the object height is assumed to be 0.26 m.

(1)   Calculate the minimum curve length required in order to satisfy the requirements of minimum sight stopping distance
(2)   Recalculate the minimum curve length with the object height assumed to be zero.

Solution (1)

$$p = +0.04$$
$$q = -0.025$$

From Table 6.7 the desirable minimum stopping distance is 160 m.

$$e = -(q - p)\frac{L}{8} = -(-0.025 - 0.04) \times 160 \div 8$$
$$= 1.3\,\mathrm{m}$$

Since $e > H_1$, $S \leq L$ as the sight distance is contained within the curve length. Therefore, using Equation 6.49:

$$L_{\mathrm{m}} = \frac{AS^2}{[\sqrt{2H_1} + \sqrt{2H_2}]^2} = \frac{0.065 \times 160^2}{[\sqrt{2 \times 1.05} + \sqrt{2 \times 0.26}]^2} = 353\,\mathrm{m}$$

Solution (2)

If the object height is assumed to be zero, then Equation 6.49 reduces to Equation 6.58:

Contd

Example 6.8 Contd

$$L = \frac{AS^2}{2H_1} = \frac{0.065 \times 160^2}{2 \times 1.05}$$

$$= 792 \text{ m}$$

Thus the required crest curve length more than doubles in value if the object height is reduced to zero.

Example 6.9

Using the same basic data as Example 6.8, but with the following straight-line gradients:

p = +0.02
q = −0.02

calculate the required curve length assuming a motorist's eye height of 1.05 m and an object height of 0.26 m.

Solution

In this case:

$$e = -(q-p)\frac{L}{8} = -(-0.02 - 0.02) \times 160 \div 8$$

$$= 0.8 \text{ m}$$

Given that in this case $e < H_1$, $S > L$ as the sight distance is greater than the curve length.
Therefore, using Equation 6.49:

$$L_m = 2S - \frac{2[\sqrt{H_1} + \sqrt{H_2}]^2}{A} = 2 \times 160 - \frac{2[\sqrt{1.05} + \sqrt{0.26}]^2}{0.04}$$

$$= 320 - 117.75$$

$$= 202.25 \text{ m}$$

Note: if the object height is reduced to zero, then the required curve length is calculated from Equation 6.64:

$$L = 2S - \frac{2H_1}{A} = 2 \times 160 - \frac{2 \times 1.05}{0.04}$$

$$= 320 - 52.5$$

$$= 267.5 \text{ m}$$

6.6.7   *Vertical sag curve design and sight distance requirements*

In general, the two main criteria used as a basis for designing vertical sag curves are driver comfort and clearance from structures.

Driver comfort

Although it is conceivable that both crest and sag curves can be designed on the basis of comfort rather than safety, it can be generally assumed that, for crest curves, the safety criterion will prevail and sight distance requirements will remain of paramount importance. However, because of the greater ease of visibility associated with sag curves, comfort is more likely to be the primary design criterion for them.

Where comfort is taken as the main criterion, the following formula is utilised in order to calculate the required curve length:

$$L = \frac{V^2 A}{3.9} \qquad (6.66)$$

where
L is the required vertical sag curve length (m)
V is the speed of the vehicle (km/hr)
A is the algebraic difference in the straight-line gradients

The vertical radial acceleration of the vehicle is assumed to be $0.3\,\text{m/s}^2$ within Equation 6.66.

Clearance from structures

In certain situations where structures such as bridges are situated on sag curves, the primary design criterion for designing the curve itself may be the provision of necessary clearance in order to maintain the driver's line of sight.

Commercial vehicles, with assumed driver eye heights of approximately 2 m, are generally taken for line of sight purposes, with object heights again taken as 0.26 m.

Again, as with crest curves, two forms of the necessary formula exist, depending on whether the sight distance is or is not contained within the curve length.

Case (1) $S \leq L$

$$L_\text{m} = \frac{A S^2}{8[C1 - (H_1 + H_2)/2]} \qquad (6.67)$$

where
Cl is the clearance height on the relevant structure located on the sag curve, generally taken in ideal circumstances at 5.7 m for bridge structures.

A, H_1, H_2 and S are as above.

Case (2) S > L

$$L_m = 2S - \frac{8[C1 - (H_1 + H_2)/2]}{A} \tag{6.68}$$

Example 6.10

A highway with a design speed of 100 km/hr is designed with a sag curve connecting a descending gradient of 3% with an ascending gradient of 5%.

(1) If comfort is the primary design criterion, assuming a vertical radial acceleration of 0.3 m/s², calculate the required length of the sag curve (comfort criterion).
(2) If a bridge structure were to be located within the sag curve, with a required clearance height of 5.7 m, then assuming a driver's eye height of 2 m and an object height of 0.26 m, calculate the required length of the sag curve (clearance criterion).

Solution (1)

$$L = \frac{V^2 A}{3.9} = \frac{100^2 \times 0.08}{3.9}$$
$$= 205\,m$$

Solution (2)

The design speed of 100 km/hr gives a desired sight stopping distance of 215 m

$$e = -(q - p)\frac{L}{8} = -(-0.05 - 0.03) \times 215 \div 8$$
$$= 2.15\,m, \text{ which is greater than the driver's eye height of 2 m}$$

Since $e > H_1$, $S < L$ as the sight distance lies outside the curve length. Thus, utilising Equation 6.67:

$$L_m = \frac{AS^2}{8[C1 - (H_1 + H_2)/2]}$$
$$= \frac{0.08 \times 215^2}{8[5.7 - (2.0 + 0.26)/2]}$$
$$= 101\,m$$

Sag curves in night-time conditions

A critical design concern for sag curves during night-time conditions can be headlight sight distance, where the length of the highway illuminated by the car's headlights is the governing parameter. The critical measurement in this instance will be the height of the headlights above the surface of the highway. This process is, however, highly sensitive to the angle of upward divergence of the light beam.

The governing formulae are:

$$L_{night} = \frac{AS^2}{2[H1 + S\tan\beta]} \qquad \text{for } S \le L \qquad (6.69)$$

$$= 2S - \frac{2[H1 + S\tan\beta]}{A} \qquad \text{for } S > L \qquad (6.70)$$

where

H1 is the height of the headlight above the highway in metres, normally assumed as 0.61 m.

S is the required sight stopping distance in metres, dependent on design speed

β is the inclined upward angle of the headlight beam relative to the horizontal plane of the vehicle (in degrees).

6.7   References

Bannister, A. & Raymond, S. (1984) *Surveying*. Longman Scientific and Technical, Harlow, Essex, UK.

DoT (1993) Highway Link Design. Departmental Standard TD 9/93. *Design Manual for Roads and Bridges, Volume 6, Road Geometry*. The Stationery Office, UK.

DoT (1996) Cross-sections and Headrooms. Departmental Standard TD 27/96. *Design Manual for Roads and Bridges, Volume 6*, Road Geometry. The Stationery Office, London, UK.

O'Flaherty, C. (1986) *Highways: Traffic Planning and Engineering*, Vol. 1. Edward Arnold, London.

Chapter 7

Highway Pavement Materials and Design

7.1 Introduction

A highway pavement is composed of a system of overlaid strata of chosen processed materials that is positioned on the in-situ soil, termed the subgrade. Its basic requirement is the provision of a uniform skid-resistant running surface with adequate life and requiring minimum maintenance. The chief structural purpose of the pavement is the support of vehicle wheel loads applied to the carriageway and the distribution of them to the subgrade immediately underneath. If the road is in cut, the subgrade will consist of the in-situ soil. If it is constructed on fill, the top layers of the embankment structure are collectively termed the subgrade.

The pavement designer must develop the most economical combination of layers that will guarantee adequate dispersion of the incident wheel stresses so that each layer in the pavement does not become overstressed during the design life of the highway.

The major variables in the design of a highway pavement are:

- The thickness of each layer in the pavement
- The material contained within each layer of the pavement
- The type of vehicles in the traffic stream
- The volume of traffic predicted to use the highway over its design life
- The strength of the underlying subgrade soil.

There are three basic components of the highway pavement, general definitions of which are given here. (More detailed descriptions of their composition appear in the explanations of the two major pavement types later in the chapter.)

Foundation

The foundation consists of the native subgrade soil and the layer of graded stone (subbase and possibly capping) immediately overlaying it. The function of the subbase and capping is to provide a platform on which to place the roadbase material as well as to insulate the subgrade below it against the effects of inclement weather. These layers may form the temporary road surface used during the construction phase of the highway.

Roadbase

The roadbase is the main structural layer whose main function is to withstand the applied wheel stresses and strains incident on it and distribute them in such a manner that the materials beneath it do not become overloaded.

Surfacing

The surfacing combines good riding quality with adequate skidding resistance, while also minimising the probability of water infiltrating the pavement with consequent surface cracks. Texture and durability are vital requirements of a good pavement surface as are surface regularity and flexibility.

For flexible pavements, the surfacing is normally applied in two layers – basecourse and wearing course – with the basecourse an extension of the roadbase layer but providing a regulating course on which the final layer is applied.

In the case of rigid pavements, the structural function of both the roadbase and surfacing layers are integrated within the concrete slab.

In broad terms, the two main pavement types can be described briefly as:

- *Flexible pavements* The surfacing and roadbase materials, bound with bitumen binder, overlay granular unbound or cement-bound material.
- *Rigid pavements* Pavement quality concrete, used for the combined surfacing and roadbase, overlays granular cement-bound material. The concrete may be reinforced with steel.

The general layout of these two pavement types is shown in Figs 7.1 and 7.2.

Pavements are thus composed of several layers of material. They can consist of one or more bitumen or cement-bound layers overlaying one or more layers of unbound granular material which in turn is laid on the in-situ soil (if the highway is in cut) or imported soil/granular material (if the highway is constructed in fill) which exists below formation level (HD 23/99) (DoT, 1999).

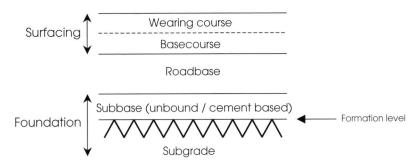

Figure 7.1 Layers within a typical flexible highway pavement.

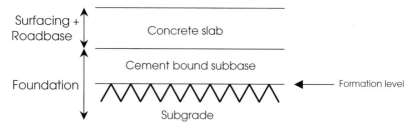

Figure 7.2 Layers within a typical rigid highway pavement.

7.2   Soils at subformation level

7.2.1   *General*

Unless the subsoil is composed of rock, it is unlikely to be strong enough to carry even construction traffic. Therefore it is necessary to superimpose additional layers of material in order to reduce the stresses incident on it due to traffic loading.

The in-situ soil would suffer permanent deformation if subjected to the high stresses arising from heavy vehicle traffic loading. The shear strength and stiffness modulus are accepted indicators of the susceptibility of the soil to permanent deformation. A soil with high values of both these characteristics will be less susceptible to permanent deformation. Both are usually reduced by increases in moisture content. Knowledge of them is essential within the pavement design process in order to determine the required thickness of the pavement layers.

Since it is not always feasible to establish these two parameters for a soil, the California bearing ratio (CBR) test is often used as an index test. While it is not a direct measure of either the stiffness modulus or the shear strength, it is a widely used indicator due to the level of knowledge and experience with it that has been developed by practitioners.

7.2.2   *CBR test*

The CBR test acts as an attempt to quantify the behavioural characteristics of a soil trying to resist deformation when subject to a locally applied force such as a wheel load. Developed in California before World War II, to this day it forms the basis for the pre-eminent empirical pavement design methodology used in the UK.

The test does not measure any fundamental strength characteristic of the soil. It involves a cylindrical plunger being driven into a soil at a standard rate of penetration, with the level of resistance of the soil to this penetrative effort being

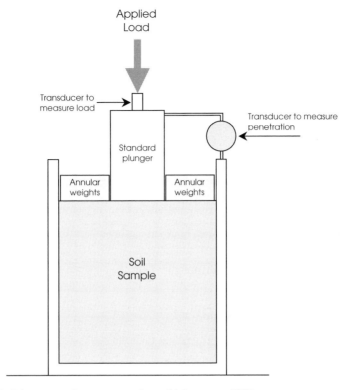

Figure 7.3 Diagrammatic representation of laboratory CBR apparatus.

measured. The test can be done either on site or in the laboratory. A diagrammatic representation of the laboratory apparatus is given in Fig. 7.3.

If the test is done in the laboratory, it is important that the moisture content and dry density of the sample being tested should approximate as closely as possible those expected once the pavement is in place. All particles greater than 20 mm in diameter should first be removed. If done in situ, the test should be performed on a newly exposed soil surface at such a depth that seasonal variations in moisture content would not be expected (see BS 1377) (BSI, 1990a).

At the start of the test, the plunger is seated under a force of 50 N for a soil with an expected CBR value of up to 30% or 250 N for an expected CBR greater than this. It then proceeds to penetrate the soil specimen at a uniform rate of 1 mm per minute. For every 0.25 mm of penetration, up to a maximum of 7.5 mm, the required loading is noted.

A graph of force versus penetration is plotted and a smooth curve drawn through the relevant points. These values are compared against the standard force-penetration relationship for a soil with a 100% CBR, the values for which are given in Table 7.1.

The CBR is estimated at penetrations of 2.5 mm and 5 mm. The higher of the two values is taken.

Penetration (mm)	Load (kN)
2	11.5
4	17.6
6	22.2
8	26.3
10	30.3
12	33.5

Table 7.1 Standard force-penetration relationship (CBR = 100%)

Example 7.1

A CBR test on a sample of subgrade yielded the data shown in Table 7.2.

Penetration (millimetres)	Load (kN)
0.5	1.6
1.0	3.3
1.5	4.9
2.0	6.6
2.5	8.2
3.0	9.3
3.5	10.5
4.0	11.4
4.5	12.2
5.0	13.0

Table 7.2 Laboratory CBR results of sample

Determine the CBR of the subgrade.

Solution

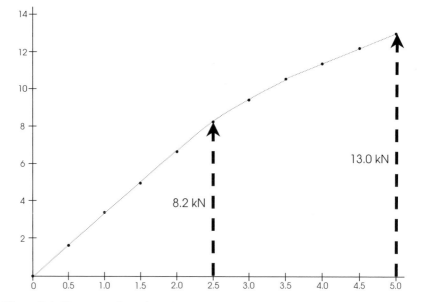

Figure 7.4 CBR curve for subgrade sample tested.

Contd

Example 7.1 Contd

At 2.5 mm penetration:

Soil = 8.2 kN

Aggregate with 100% CBR = 13.02 kN

Therefore

CBR = $(8.2 \times 100) \div 13.02$

= 63%

At 5.0 mm penetration:

Soil = 13.0 kN

Aggregate with 100% CBR = 19.9 kN

Therefore

CBR = $(13.0 \times 100) \div 19.9$

= 65.3%

Taking the larger of the two values:

Final CBR = 65.3% → 65%

Note: CBR values are rounded off as follows:

CBR ≤ 30% – round to nearest 1%

CBR > 30% – round to nearest 5%

7.2.3   Determination of CBR using plasticity index

Where it is not possible to determine the CBR of a given soil directly, an alternative methodology involving use of the soils plasticity index and a knowledge of certain service conditions can be used to derive a CBR valuation for cohesive soils (Black & Lister, 1979).

In order to derive the plasticity index of a soil, its liquid and plastic limit must be obtained.

Liquid limit

The liquid limit is the moisture content at which the soil in question passes from the plastic to the liquid state. It is derived using the cone penetrometer test. In it, a needle of a set shape and weight is applied to the surface of a soil sample placed in a standard metal cup and allowed to bear on it for a total of 5 seconds.

The penetration of the needle into the sample is measured to the nearest tenth of a millimetre. The moisture content of the sample is then determined.

The process is repeated four more times, on each occasion with a sample of differing moisture content. A relationship between cone penetration and moisture content can then be established, allowing the moisture content corresponding to a cone penetration of 20 mm to be determined. This moisture content is termed the liquid limit of the soil under examination. See BS 1377 for further details of the cone penetrometer test.

Plastic limit

The plastic limit is defined as the moisture content at which the soil in question becomes too dry to be in a plastic condition. The plastic limit test, as defined by BS 1377, involves taking a 15 g soil sample, mixing it with water, and rolling it into a 3 mm diameter thread. (The rolling process will reduce the moisture content of the sample.) This process is done repeatedly for different samples until the point is reached when the sample just crumples when rolled into a 3 mm diameter thread. The moisture content of the sample in question can be taken as the plastic limit of that soil.

Plasticity index

The plasticity index of a soil is defined as the liquid limit of a soil minus its plastic limit:

Plasticity index (PI) = Liquid limit (LL) − Plastic limit (PL)   (7.1)

It denotes the moisture content range over which the soil is in a plastic state.

Using plasticity index to derive CBR

If it is not possible to derive the CBR of a soil using the standard test referred to in section 7.2.2, its plasticity index can be used as a means of assessing it (Black & Lister, 1979). This method determines the long-term CBR of various subgrades, as shown in Table 7.3.

Notes to Table 7.3:

(1)   A high water table is one situated less than 300 mm below formation level
(2)   A low water table is one situated more than 1 m below formation level
(3)   Poor conditions denote the situation where the lowest layer of the pavement is laid on weak soil in heavy rain
(4)   Average conditions denote the situation where the formation is protected during adverse weather
(5)   Good conditions denote the situation where the soil is drier than its likely service conditions during construction

Table 7.3 CBR values for different soil types and conditions

| | | High water table | | | | | | Low water table | | | | | |
| | | Poor | | Average | | Good | | Poor | | Average | | Good | |
	PI	A	B	A	B	A	B	A	B	A	B	A	B
Heavy clay	70	1.5	2	2	2	2	2	1.5	2	2	2	2	2.5
	60	1.5	2	2	2	2	2.5	1.5	2	2	2	2	2.5
	50	1.5	2	2	2.5	2	2.5	2	2	2	2.5	2	2.5
	40	2	2.5	2.5	3	2.5	3	2.5	2.5	3	3	3	3.5
Silty clay	30	2.5	3.5	3	4	3.5	5	3	3.5	4	4	4	6
Sandy clay	20	2.5	4	4	5	4.5	7	3	4	5	6	6	8
	10	1.5	3.5	3	6	3.5	7	2.5	4	4.5	7	6	>8
Silt	—	1	1	1	1	2	2	1	1	2	2	2	2
Sand													
Poorly graded	—	—	—	—	—	—	20	—	—	—	—	—	—
Well graded	—	—	—	—	—	—	40	—	—	—	—	—	—
Sandy gravel	—	—	—	—	—	—	60	—	—	—	—	—	—

Table 7.4 CBR estimates where information is poor

Soil type	PI	CBR (%)
Heavy clay	70	2
	60	2
	50	2
	40	2/3
Silty clay	30	3/4
	20	4/5
Sandy clay	10	4/5
Sand		
Poorly graded	—	20
Well graded	—	40
Sandy gravel		
Well graded	—	60

(6) 'A' denotes the situation where the pavement is 300 mm thick (thin pavement construction)

(7) 'B' denotes the situation where the pavement is 1.2 m thick (thick pavement construction).

If full information is not available for Table 7.3, certain assumptions can be made. The worst service condition of 'high water table' can be assumed, together with the assertion that construction is being carried out in accordance with standard specifications, taken as 'average' construction conditions in Table 7.3. If the pavement thickness varies between the two values of 300 mm and 1.2 m, the final CBR can be derived by interpolation between the values given in Table 7.3.

Where full information is unavailable, general CBR values of the type given in Table 7.4 can be used (HD 25/94) (DoT, 1994).

7.3 Subbase and capping

7.3.1 General

The subbase and capping together act as a regulator of the surface of the sub-grade below and protect it against the effects of inclement weather. They, along with the subgrade, provide a secure platform on which the upper layers of the highway pavement can be built.

The determinant of the thickness of this section of the pavement is the strength of the underlying subgrade. Its design is independent of the cumulative traffic incident on the upper layers of the pavement over its design life. For subgrades in excess of 5% CBR, the required subbase depth is no greater than 225 mm, down to a minimum of 150 mm at a subgrade CBR of 15% (HD 25/94).

Granular and cement-based subbases are recommended for flexible pavements while only cemented subbases are recommended for rigid-type pavements (HD 25/94). In the case of unbound subbases, their grading should be such that it constitutes a dense layer of relatively high stiffness modulus, relatively imper-meable to water though not of necessity free draining. Their laboratory CBR should be a minimum of 30%.

7.3.2 Thickness design

The thickness of the subbase/capping layer is dependent on the CBR of the subgrade and is determined in accordance with HD 25/94 using Fig. 7.5.

Figure 7.5 illustrates two separate designs, one with subbase only where capping is not required (denoted by the heavy dotted line) and one comprising subbase combined with capping (denoted by the heavy continuous line).

The following four broad categories apply:

(1) No subbase is required if the subgrade is composed of hard rock or of a granular material with a CBR of at least 30%, provided the water table is not at a high level.

(2) In the case of subgrades with a CBR greater than 15%, a subbase thick-ness of 150 mm is required (in practical terms this constitutes the minimum subbase thickness for proper spreading and compaction).

(3) Where the CBR of the subgrade lies between 2.5% and 15%, two options are available:

(a) use 150 mm of subbase over a layer of capping material, the thick-ness of which depends on the subgrade CBR, *or*

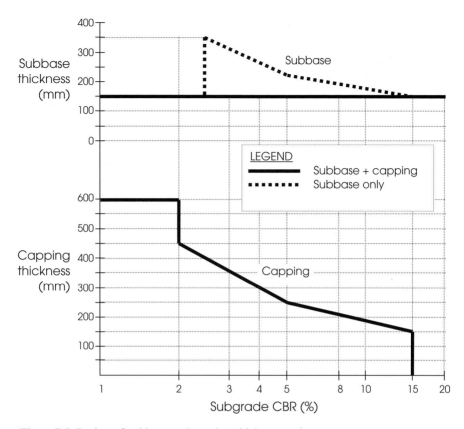

Figure 7.5 Design of subbase and capping thickness.

    (b)   a layer of subbase varying between 150 mm (at 15% CBR) and 350 mm (at 2.5% CBR) in thickness.

(4)   For all pavements where the subgrade CBR is below 2.5% and for rigid pavement construction on materials with CBR less than 15%, 150 mm of subbase must be used on top of capping. The thickness of the capping layer will reach 600 mm where the CBR of the subgrade dips below 2%.

(5)   Where the subgrade CBR is substantially below 2%, the material will often be removed in favour of more suitable material. The depth of this imported material would typically be between 500 mm and 1000 mm deep. Though this material may in reality be quite strong, it will be assumed to have a CBR of 2% and will thus require a 600 mm capping layer.

7.3.3   *Grading of subbase and capping*

Type 1 granular materials are usually employed in subbases for flexible type pavements. In situations, however, where the design traffic loading for the pavement at opening is predicted to be less than 5 million standard axles, a Type 2 material can be utilised. A standard axle is defined as 80 kN with the cumulative number normally expressed in millions of standard axles (or msa). For all unbound granular subbases, the CBR must be a minimum of 30%.

Type 1 is seen as the most suitable both because of its free draining nature and the high degree of interlock it helps to develop between the aggregate particles. It is, however, more expensive. It can be crushed rock or slag or concrete with up to 12.5% by mass passing the 5 mm sieve and the fraction passing the 425 micron sieve being non-plastic. Type 2 can be sand, gravel, crushed rock or slag or concrete with the fraction passing the 425 micron sieve having a plasticity index (PI) of less than 6.

Table 7.5 gives the grading for both Type 1 and Type 2 granular materials (*Specification for Highway Works*, 1998).

BS sieve size	Percentage passing by mass	
	Type 1 subbase	Type 2 subbase
75 mm	100	100
37.5 mm	85–100	85–100
20 mm	60–100	60–100
10 mm	40–70	40–100
5 mm	25–45	25–85
0.6 mm	8–22	8–45
0.075 mm	0–10	0–10

Table 7.5 Grading requirements of subbase materials for use within flexible pavements in the UK

For rigid-type pavements, a cemented subbase is required to minimise the risk of water penetrating the slab joints and cracks and thereby weakening the subbase itself. An impermeable membrane should be placed over the subbase prior to the construction of the upper layers of the pavement. Ideally, strong cement bound material (CBM3) should be used, unless the design traffic loading at opening is less than 12 msa, where weak cement bound material (CBM2) becomes permissible.

It should be noted that weak cement bound materials (CBM1 or CBM2) can be used as subbases for flexible type pavements. CBM3, CBM4 and CBM5 are high quality materials prepared in most cases at a central plant from batched quantities of material such as crushed rock or gravel. CBM2 is usually processed from sand/gravel or crushed rock while CBM1 may include unprocessed granular soils. Both CBM1 and CBM 2 can be mixed in situ rather than at a central plant.

Typically, strengths of 7 N/mm^2 are required for cement bound materials. Lean-mix concrete can also be used as subbase material for concrete pavements. A dry lean-mix concrete would typically have a strength of 10 N/mm^2.

Table 7.6 gives the grading for cement bound materials CBM1, CBM2 and CBM3/4/5 as well as for a typical dry lean concrete (*Specification for Highway Works*, 1998)

Table 7.6 Grading requirements of cement-bound and lean concrete materials for use in subbases within both flexible and rigid pavements in the UK

	Percentage passing by mass			
BS sieve size	CBM1	CBM2	CBM3/CBM4/CBM5 (40 mm nominal max size)	Dry lean concrete
75 mm	100	100	100	100
37.5 mm	95	95–100	95–100	95–100
20 mm	45	45–100	45–80	45–80
10 mm	35	35–100	—	—
5 mm	25	25–100	25–50	25–50
2.36 mm	—	15–90	—	—
0.6 mm	8	8–65	8–30	8–30
0.3 mm	5	5–40	—	—
0.15 mm	—	—	0–8	0–8
0.075 mm	0	0–10	0–5	0

7.4 Traffic loading

When designing a new highway, the estimation of traffic levels at opening is of central importance to the structural design of the upper layers of the road pavement.

Of particular importance is the estimation of commercial vehicle volumes. Commercial vehicles are defined as those with an unladen weight of 15 kN. They are the primary cause of structural damage to the highway pavement, with the damage arising from private cars negligible in comparison.

The following is the classification for commercial vehicles used in HD 24/96 (DoT 1996):

- Buses and coaches (PSV)
- 2 axle rigid (OGV1)
- 3 axle rigid (OGV1)
- 3 axle articulated (OGV1)
- 4 axle rigid (OGV2)
- 4 axle articulated (OGV2)
- 5+ axles (OGV2).

These are illustrated graphically in Fig. 7.6.

In order to allow the determination of the cumulative design traffic for the highway in question, therefore, the total flow of commercial vehicles per day in one direction at the day of opening (or, for maintenance purposes, at the present time) plus the proportion of vehicles in the OGV2 category must be ascertained. If all flow data is two-directional, then a 50:50 split is assumed unless available data demonstrates otherwise.

Figure 7.7 is a representation of the graph detailed in HD 24/96 for estimat-

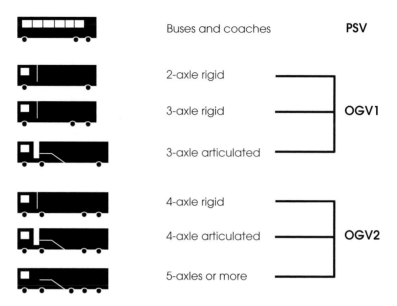

Figure 7.6 Vehicle classifications.

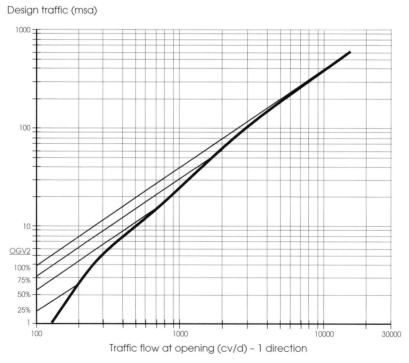

Figure 7.7 Design traffic for flexible and flexible composite pavements (20-year design life) – single carriageway (HD 24/96) (DoT, 1996).

Design traffic (msa)

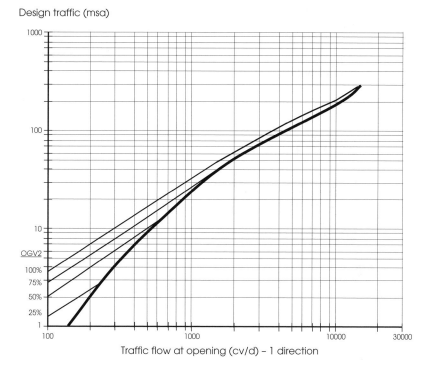

Traffic flow at opening (cv/d) – 1 direction

Figure 7.8 Design traffic for flexible and flexible composite pavements (20-year design life) – dual carriageway (HD 24/96) (DoT, 1996).

ing the cumulative design traffic over a 20-year design life for a flexible single carriageway pavement. Figure 7.8 is a representation of the graph for flexible dual carriageway pavements. Figure 7.9 represents the graph for a single carriageway rigid pavement (40-year design life). Figure 7.10 is a representation of the graph for rigid dual carriageway pavements.

The million standard axle valuation derived from these graphs includes an adjustment required to estimate the left-hand lane traffic which is subsequently used for pavement design in Chapter 8 using HD 26/01 (DoT, 2001).

While economic studies have shown that a design life of 40 years is optimal, flexible or partially flexible pavements are normally designed initially for 20 years, after which major maintenance is carried out.

Example 7.2

The one-directional commercial vehicle flow data shown in Table 7.7 was collected as the estimate of the opening day flow for a proposed new highway.

Table 7.7 Traffic count data

Commercial vehicle type	Classification	Number of vehicles
Buses/coaches	PSV	80
2 axle rigid	OGV1	480
3 axle rigid	OGV1	70
3 axle articulated	OGV1	90
4 axle rigid	OGV2	220
4 axle articulated	OGV2	280
5+ axle	OGV2	220

Solution

From the data supplied in Table 7.7, the total flow, total OGV2 flow and the percentage OGV2 flow are as follows:

Total flow: 1440 commercial vehicles
Total OGV2 flow: 720
Percentage OGV2:   50%

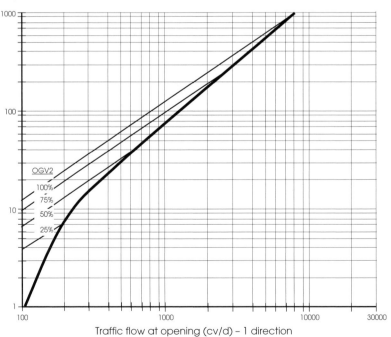

Figure 7.9 Design traffic for rigid, rigid composite and flexible pavement (40-year design life) – single carriageway (HD 24/96) (DoT, 1996).

Design traffic (msa)

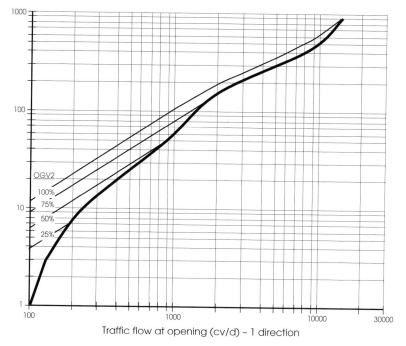

Figure 7.10 Design traffic for rigid, rigid composite and flexible pavement (40-year design life) – dual carriageway (HD 24/96) (DoT, 1996).

Example 7.3

Using the total commercial flow per day in one direction at opening and the proportion in the OGV2 category from the previous example, estimate the cumulative design traffic for the four road types detailed in Figs 7.7 to 7.10.

Solution

(1) From Fig. 7.7, for flexible and flexible composite pavements (20-year design life) – single carriageway
Cumulative design traffic = 40 million standard axles (msa)
(2) From Fig. 7.8, for flexible and flexible composite pavements (20-year design life) – dual carriageway
Cumulative design traffic = 35 million standard axles (msa)
(3) From Fig. 7.9, for rigid, rigid composite and flexible pavement (40-year design life) – single carriageway
Cumulative design traffic = 105 million standard axles (msa)
(4) From Fig. 7.10, for rigid, rigid composite and flexible pavement (40-year design life) – dual carriageway
Cumulative design traffic = 100 million standard axles (msa)

Figure 7.11 illustrates how the solutions to these four cases are derived.

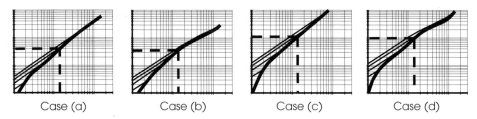

| Case (a) | Case (b) | Case (c) | Case (d) |

Figure 7.11 Derivation of cumulative design traffic valuations.

7.5   Pavement deterioration

7.5.1   *Flexible pavements*

Experience has indicated that, for heavily trafficked roads, deterioration in the form of cracking/deformation is most likely to be found in the surface of the pavement rather than deeper down within its structure. A well-constructed pavement will have an extended life span on condition that distress, seen in the form of surface cracks and ruts, is taken care of before it starts to affect the structural integrity of the highway (HD 26/01) (DoT, 2001).

There are four basic phases of structural deterioration for a flexible pavement (HD 26/01):

Phase 1
When a new/strengthened pavement is reaching stability, at which point its load spreading ability is still improving.

Phase 2
Load spreading ability is quite even and the rate of structural deterioration can be calculated with some confidence.

Phase 3
At this stage structural deterioration becomes less predictable and strength may decrease gradually or even rapidly. This is the 'investigatory' phase. A pavement entering this phase should be monitored in order to ascertain what if any remedial action is required to be carried out on it. (Residual life is defined as the period of time before a pavement reaches this phase.)

Phase 4
Here the pavement has deteriorated to failure. Strengthening can only be achieved by total reconstruction. This phase can last quite a number of years, with maintenance becoming necessary with increasing frequency until the point is reached where the costs associated with this treatment make reconstruction the cheaper option.

Remedial work during the third 'investigatory' phase is more economic than total reconstruction at the end of its full design life. If replacement of the surfacing or overlaying on top of it is expedited at the start of this phase, the time to failure can be greatly extended.

7.5.2 Rigid pavements

Cracking in rigid concrete slabs can be promoted by stresses generated at the edge/corner of slabs. These can vary from narrower hairline cracks which often appear while concrete is drying out, to 'wide' cracks (>0.15 cm) which result in the effective loss of aggregate interlock, allowing water to enter its structure and cause further deterioration. 'Medium' cracking greater than 0.5 mm will result in partial loss of aggregate interlock.

Failure is defined as having occurred in an unreinforced concrete pavement if one of the following defects is present (HD 26/01)(DoT, 2001):

- A medium or wide crack crossing the bay of the concrete slab longitudinally or transversely
- A medium longitudinal and medium transverse crack intersecting, both exceeding 200 mm in length and starting from the edge of the pavement
- A wide corner crack, more than 200 mm in radius, centred on the corner.

7.6 Materials within flexible pavements

7.6.1 Bitumen

Bitumen is produced artificially from crude oil within the petroleum refining process. It is a basic constituent of the upper layers in pavement construction. It can resist both deformation and changes in temperature. Its binding effect eliminates the loss of material from the surface of the pavement and prevents water penetrating the structure. Two basic types of bituminous binder exist:

- Tar – obtained from the production of coal gas or the manufacture of coke
- Bitumen – obtained from the oil refining process.

With the decreased availability of tar, bitumen is the most commonly used binding/water resisting material for highway pavements.

The oil refining process involves petroleum crude being distilled, with various hydrocarbons being driven off. The first stage, carried out at atmospheric pressure, involves the crude being heated to approximately 250°C. Petrol is the most volatile of these and is driven off first, followed by materials such as kerosene and gas oil. The remaining material is then heated at reduced pressure to collect the diesel and lubricating oils contained within it. At the conclusion of this stage of the process a residue remains which can be treated to produce bitumen of varying penetration grades. This is the material used to bind and stabilise the graded stone used in the top layers of a highway pavement.

A number of tests exist to ensure that a binder has the correct properties for use in the upper layers of a pavement. Two of the most prominent are the penetration test and the softening point test, both of which indirectly measure the viscosity of a sample of bitumen. (The viscosity of a fluid slows down its ability to flow and is of particular significance at high temperatures when the ability of the bitumen to be sprayed onto or mixed with aggregate material is of great significance.) The penetration test is in no way indicative of the quality of the bitumen but it does allow the material to be classified.

The penetration test involves a standard steel needle applying a vertical load of 100 g to the top of a standard sample of bitumen at a temperature of 25°C. The depth to which the needle penetrates into the sample within a period of 5 seconds is measured. The answer is recorded in units of 0.1 mm. Thus, if the needle penetrates 10 mm within the five second period the result is 100 and the sample is designated as 100 pen. The lower the penetration the more viscous and therefore the harder the sample. Figure 7.12 is a diagrammatic representation of the penetration test.

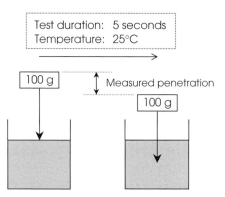

Figure 7.12
Penetration test for bitumen.

The softening point test involves taking a sample of bitumen which has been cast inside a 15 mm diameter metal ring and placing it inside a water bath with an initial temperature of 5°C. A 25 mm clear space exists below the sample. A 10 mm steel ball is placed on the sample and the temperature of the bath and the sample within it is increased by 5°C per minute. As the temperature is raised, the sample softens and therefore sags under the weight of the steel ball. The temperature at which the weakening binder reaches the bottom of the 25 mm vertical gap below its initial position is known as its softening point. An illustration of the softening point test is given in Fig. 7.13.

Bitumen should never reach its softening point while under traffic loading.

The results from these two tests enable the designer to predict the temperatures necessary to obtain the fluidity required in the mixture for effective use within the pavement.

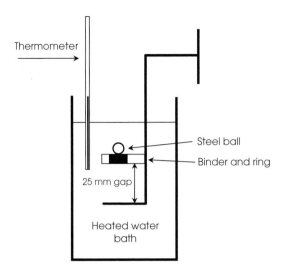

Figure 7.13 Softening point test.

Table 7.8 indicates the penetration and softening point valuations for different bitumen grades (BS 3690) (BSI, 1990b).

Table 7.8 Properties of penetration grade bitumens (BSI, 1990b)

Property	Grade of bitumen			
	15 pen	50 pen	100 pen	200 pen
Penetration at 25°C	15 ± 5	50 ± 10	100 ± 20	200 ± 30
Softening point (°C) min.	63	47	41	33
Softening point (°C) max.	76	58	51	42

7.6.2 Surface dressing and modified binders

Surface dressing involves the application of a thin layer of bituminous binder to the surface of the pavement slab followed by the spreading and rolling into it of single sized stone chippings. In order to apply the binder effectively, its stiffness must be modified during the construction phase of the pavement. Two such binder modifications used during surface dressing are cutback bitumen and bitumen emulsion.

Cutback bitumen

Bitumen obtained from the refining process described briefly above can be blended with some of the more volatile solvents such as kerosene or creosote to form a solution that has a viscosity far below that of penetration grade bitumen and will act as a fluid at much lower temperatures. However, when the solution is exposed to the atmosphere, the volatile solvents evaporate leaving solely the bitumen in place. Such solutions are termed cutbacks and the process of evaporation of the volatile solvents is called curing. The speed at which it occurs will depend on the nature of the solvent.

The classification of cutbacks is based on the following two characteristics:

- The viscosity of the cutback itself
- The penetration of the non-volatile residue.

The cutback's viscosity is measured using a standard tar viscometer (STV) which computes the time in seconds for a given volume of binder to flow through a standard orifice at a temperature of 40°C. Three common grades for cutback have viscosities of 50, 100 and 200 seconds. Cutback bitumen is used in surface dressing. In this process, it is sprayed onto a weakened road surface and chippings are placed on it and then rolled. It serves to provide a non-skid wearing surface to the pavement, makes the surface resistant to water and prevents its disintegration.

Bituminous emulsions

Bitumen can be made easier to handle by forming it into an emulsion where particles of it become suspended in water. In most cases, their manufacture involves heating the bitumen and then shredding it in a colloidal mill with a solution of hot water and an emulsifier. The particles are imparted with an ionic charge which makes them repel each other. Within cationic emulsions the imparted charge is positive, while the charge is negative in anionic emulsions. When the emulsion is sprayed onto the road surface, the charged ions are attracted to opposite charges on the surface, causing the emulsion to begin 'breaking' with the bitumen particles starting to coalesce together. The breaking process is complete when the film of bitumen is continuous.

Bitumen emulsions are graded in terms of their stability or rate of break on a scale of 1 to 4, with 1 signifying the greatest stability (stable = rapid acting). Rate of break depends on the composition of the emulsion and the rate at which the emulsion evaporates. The grading of the aggregate onto which the emulsion is applied is also important to the rate of break. Dirty aggregates accelerate it, as will porous or dry road surfaces. Cationic emulsions tend to break more rapidly than ionic ones. The UK code, BS 434 (BSI, 1984), also designates cationic emulsions as K and ionic as A. Therefore, K3 denotes a slow acting cationic emulsion, K2 a medium acting one and K1 a rapid acting one.

Chippings

The chippings used are central to the success of the surface dressing process as they provide essential skidding resistance. The correct rate of spread depends mainly on the nominal size of chippings used, varying from 7 kg per m² for 6 mm nominal size to 17 kg per m² for 20 mm. The chippings themselves may be precoated with a thin layer of binder in order to promote their swift adhesion to the binder film during the laying process. Rolling should be carried out using pneumatic-tyred rollers. The process should result in a single layer of chippings covering the entire surface, firmly held within the binder film.

7.6.3   *Recipe specifications*

Some of the most important bituminous materials used within highway pavements in the UK are:

- Coated macadam (dense bitumen macadam, high density macadam, pervious macadam)
- Asphalt (mastic asphalt, hot rolled asphalt).

The main uses for these materials within a highway pavement are shown in Table 7.9.

Bituminous material	Location in pavement
Dense bitumen macadam	Roadbase, basecourse, wearing course
High density macadam	Roadbase, basecourse
Pervious macadam	Wearing course
Mastic asphalt	Wearing course
Hot rolled asphalt	Roadbase, basecourse, wearing course

Table 7.9 Location in pavement of different bituminous materials

Hot rolled asphalt, dense bitumen macadam and porous macadam are the most prominent recipe-based bituminous materials used in major highways. The recipe method uses a cookbook-type procedure for the selection of the type and relative proportions of the materials within the mixture. This selection is based on both experience over many years and empirical judgement rather than strict theoretical engineering principles.

It involves the specification of the type of aggregate together with its grading, the grade of the bitumen and the relative proportions of the bitumen and aggregate. The method of mixing, placement and compaction will also be stipulated. This mixture is specified on the basis that it has been adjudged by experts within the industry to have performed to an acceptable level over a time span of years. It is the method that is concentrated on within this text. It does however have some limitations, most notably its inability to allow for the inclusion of more innovative road materials or to provide a workable specification where unusual traffic or climatic conditions may prevail. Furthermore, it may, in certain situations, prove impossible to ensure that the mixture has been produced exactly as the specification requires. The method, by its very nature, cannot take full account of the engineering properties of the mixture. The engineering design approach to bituminous surfacings was put forward and details of it be found elsewhere (O'Flaherty, 2002).

7.6.4   Coated macadams

With these pavement materials, graded aggregate is coated with bituminous binder, generally penetration grade bitumen. It is classified in terms of the nominal size of the aggregate, its grading and the location within the pavement for which it is intended. Densely graded materials have a high proportion of fines producing dense and stable macadam. Open graded materials have less fines. This results in less dense and less stable macadam.

Let us examine the three coated macadams used within the roadbase/base-course/wearing courses of flexible pavements – dense bitumen macadam (DBM), heavy duty macadam (HDM) and porous macadam.

Dense bitumen macadam

Dense bitumen macadam is well graded and is the most common material used in the roadbase and basecourse of major roadways (trunk roads/motorways) within the UK. Table 7.10 illustrates the properties of two types of dense bitumen macadam – 28 mm and 40 mm nominal size.

BS sieve size	Dense bitumen macadam	
	28	40
37.5 mm	100–100	95–100
28 mm	90–100	70–94
20 mm	71–95	—
14 mm	58–82	56–76
6.3 mm	44–60	44–60
3.35 mm	32–46	32–46
0.3 mm	7–21	7–21
0.075 mm	2–9	2–9
Bitumen content (% by mass of total)	3.4–4.6	2.9–4.1
Grade of binder (pen)	100	100

Table 7.10 Dense bitumen nacadam (DBM) compositions (*Specification for Highway Works*, 1998)

Heavy duty macadam

Heavy duty macadam is used in roadbases and basecourses for major highways with high traffic loadings. It contains more of the finer material (filler) and uses a harder bitumen grading than DBM. The result is a stiffer mixture that will provide greater protection against cracking and deformation over the life of the pavement. Table 7.11 illustrates the properties of two types of heavy duty macadam – 28 mm and 40 mm nominal size.

BS sieve size	Heavy duty macadam	
	28	40
37.5 mm	100	95–100
28 mm	90–100	70–94
20 mm	71–95	—
14 mm	58–82	56–76
6.3 mm	44–60	44–60
3.35 mm	32–46	32–46
0.3 mm	7–21	7–21
0.075 mm	7–11	7–11
Bitumen content (% by mass of total)	3.4–4.6	2.9–4.1
Grade of binder (pen)	50	50

Table 7.11 Heavy duty macadam (DBM) compositions (*Specification for Highway Works*, 1998)

Porous macadam

Known as porous asphalt, unlike conventional bituminous materials which provide an impermeable layer and protect the underlying layers from the ingress of rainwater, porous macadam is an open graded material containing a high proportion of voids whose primary function is to allow the rapid drainage of water. The impervious nature of the lower layers together with the camber of the road allows the rainwater to flow laterally through the porous asphalt, thereby escaping quickly from the structure.

This type of surfacing greatly improves wheel grip on the road while also reducing water spray and substantially reducing the general noise levels emanating from a highway. Due to its high voids content, this material is not as durable as the more impervious macadams and should not be used in areas of particularly high traffic loading. The bitumen in the mix can be stiffened by the addition of hydrated lime in order to reduce the likelihood of the binder being stripped away from the aggregate. A relatively high bitumen content should be employed. Table 7.12 illustrates the properties of 20 mm nominal size porous asphalt.

BS sieve size	20 mm porous asphalt
28 mm	100
20 mm	100–95
14 mm	75–55
6.3 mm	30–20
3.35 mm	13–7
0.075 mm	5.5–3.5
Bitumen content (% by mass of total)	3.4–4.5
Grade of binder (pen)	100–200

Table 7.12 Pervious macadam/porous asphalt (PA) composition (*Specification for Highway Works*, 1998)

7.6.5   *Asphalts*

Two asphalts are discussed within this section: hot rolled asphalt (HRA) and mastic asphalt.

Hot rolled asphalt

Hot rolled asphalt is similar to a coated macadam. It is a dense material with low air voids content, consisting of a mixture of aggregate, fines, binder and a filler material, but in this case the grading is far less continuous (gap-graded) with a higher proportion of both fines and binder present in the mix. The material is practically impervious to water, with the fines, filler and bitumen forming a mortar in which coarse aggregate is scattered in order to increase its overall bulk.

Hot rolled asphalt wearing courses typically have from zero to 55% coarse aggregate content, with basecourses having either 50% or 60% and roadbases normally at 60%.

There are two recipe mixes for gap-graded rolled asphalt wearing course: Type F, characterised by the use of sand fines, and Type C, characterised by the use of crushed rock or slag fines. F denotes a finer grading of the fine aggregate with C denoting a coarser grading of the fine aggregate.

Table 7.13 details a range of mixes for hot rolled asphalt to be used at roadbase, basecourse and wearing course levels within the pavement. Each mix has a designation composed of two numbers, with the first relating to the percentage coarse aggregate content in the mix and the second to the nominal coarse aggregate size. (The wearing courses in Table 7.13 are both Type F.)

As hot rolled asphalt wearing course is a smooth-textured material, precoated chippings should be spread over and rolled into its surface while plastic in order to increase skid resistance.

Mastic asphalt

Mastic asphalt is a very durable heavy-duty, weather-proof wearing course material. It consists of a mixture of asphaltic cement (low-penetration grade bitumen), fine aggregate and filler in proportions which result in a low-void impermeable mass. It contains a low percentage of coarse aggregate, all of which must pass the 14 mm sieve and be retained on the 10 mm sieve. The mix consists of a high percentage of fine aggregate, with no less than 45% and no more than 55% passing the 0.075 mm sieve and at least 97% passing the 2.36 mm sieve in addition to high proportions of both filler material and binder. The grade of the binder is very high (10 to 25 pen) with less than 1% voids in the mix.

The mix is applied manually using wooden floats, at a temperature of between 175°C and 225°C approximately. It requires considerable working, with the finished layer measuring between 40 mm and 50 mm. Its low skidding resistance

Table 7.13 Examples of hot rolled asphalt roadbase, basecourse and wearing course bituminous mixes (BS 594) (BSI, 1992)

Location	Roadbase	Basecourse	Wearing course	
Designation	60/20	50/14	30/14	30/10
BS sieve size grading				
28 mm	100	—	—	—
20 mm	90–00	100	100	—
14 mm	30–65	90–100	85–100	100
10 mm	—	65–100	60–90	85–100
6.3 mm	—	—	—	60–90
2.36 mm	30–44	35–55	60–72	60–72
0.600 mm	10–44	15–55	45–72	45–72
0.212 mm	3–25	5–30	15–50	15–50
0.075 mm	2–8	2–9	8–12	8–12
Bitumen content				
(% by mass of total mixture)				
Crushed rock/steel slag	5.7	6.5	7.8	7.8
Gravel	5.5	6.3	7.5	7.5
Blast furnace slag of bulk density				
1440 kg/m^3	5.7	6.6	7.9	7.9
1280 kg/m^3	6.0	6.8	8.1	8.1
1200 kg/m^3	6.1	6.9	8.2	8.2
1120 kg/m^3	6.3	7.1	8.3	8.3
Layer depth (mm)	45–80	25–50	40	35

requires the application of precoated chippings to its surface while still plastic, in order to embed them firmly into the surface of the mix and give a roughened finish.

It results in a long-life, low-maintenance surfacing in highly trafficked predominantly urban locations. The labour intensive nature of its application makes it costly relative to other bituminous wearing courses.

7.6.6   Aggregates

The maximum nominal aggregate size is determined from both the required thickness of the material when put in place and the surface texture called for. The following are typical nominal aggregate sizes used at different levels within a bituminous pavement:

- Wearing course
  - 14 mm dense wearing course macadam
  - 10 mm or 6 mm pervious macadam
- Base course
  - 40, 28 or 20 mm dense macadam

- Roadbase
  - ○ 40 or 28 mm dense macadam.

The size of aggregate must not be greater than the required layer thickness. The layer thickness should be approximately $2^{1}/_{2}$ times the nominal maximum aggregate size, with a minimum layer thickness of $1^{1}/_{2}$ times the nominal maximum aggregate size in order to minimise the likelihood of the larger stones being crushed during rolling.

7.6.7   *Construction of bituminous road surfacings*

The production of a successful bituminous road surfacing depends not just on the design of the individual constituent layers but also on the correctness of the construction procedure employed to put them in place. In essence, the construction of a bituminous pavement consists of the flowing steps:

- Transporting and placing the bituminous material

- Compaction of the mixture

- If required, the spreading and rolling of coated chippings into the surface of the material.

Transporting and placing

The bituminous material is manufactured at a central batching plant where, after the mixing of its constituents, the material is discharged into a truck or trailer for transportation to its final destination. The transporters must have metallic beds sprayed with an appropriate material to prevent the mixture sticking to it. The vehicle should be designed to avoid heat loss which may result in a decrease in temperature of the material, leading to difficulties in its subsequent placement – if it is too cold it may prove impossible to compact properly.

It is very important that the receiving surface is clean and free of any foreign materials. It must, therefore, be swept clean of all loose dirt. If the receiving layer is unbound, it is usual to apply a prime coat, in most cases cutback bitumen, before placing the new bituminous layer. A minimum ambient temperature of at least 4°C is generally required, with BS 594 stating that a wearing course should not be laid when the temperature of the course being covered is less than 5°C, and work stops completely when the air temperature hits 0°C on a falling thermometer. Work may, however, recommence if the air temperature hits –1°C on a rising thermometer, provided the surface is ice-free and dry.

Steps must be taken to ensure that the surface being covered is regular. If it is irregular, it will not be possible to attain a sufficiently regular finished surface. A typical surface tolerance for a bituminous basecourse or wearing course would be ±6 mm.

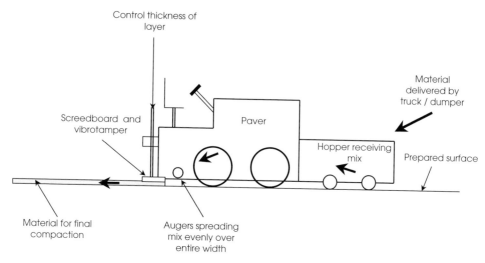

Control thickness of
layer

Material
delivered by
truck / dumper

Screedboard and
vibrotamper

Paver

Hopper receiving
mix

Prepared surface

Material for final
compaction

Augers spreading
mix evenly over
entire width

Figure 7.14 Operational features of a paving machine.

A paver (Fig. 7.14) is used for the actual placing of the bituminous material. It ensures a uniform rate of spread of correctly mixed material. The truck/trailer tips the mixture into a hopper located at the front of the paver. The mix is then fed towards the far end of the machine where it is spread and agitated in order to provide an even spread of the material over the entire width being paved. The oscillating/vibrating screed and vibrotamper delivers the mix at the required elevation and cross-section and uses a tamping mechanism to initiate the compaction process.

Compaction of the bituminous mix

When the initial placing of the mix is complete it must be rolled while still hot. Minimum temperatures vary from 75°C to 90°C depending on the stiffness of the binder. This process is completed using either pneumatic tyre or steel wheel rollers. The tyre pressures for pneumatic rollers vary from 276 kPa to 620 kPa, while the steel wheel rollers vary from 8 to 18 t. If the latter are vibratory rather than static, 50 vibrations per second will be imparted. The rolling is carried out in a longitudinal direction, generally commencing at the edge of the new surface and progressing towards the centre. (If the road is superelevated, rolling commences on the low side and progresses towards the highest point.)

It is important that, on completion of the compacting process, the surface of the pavement is sufficiently regular. Regularity in the transverse direction is measured using a simple 3-metre long straight edge. Deviations measured under the straight edge should in no circumstances exceed 3 mm.

Application of coated chippings to smooth surfacings

Chippings are frequently used in order to give improved surface texture to smooth wearing course mixes such as hot rolled asphalt. They are placed after laying but prior to compaction. The two major considerations are the uniformity and rate of spread of the chippings and the depth of their embedment – deep enough so that the bituminous mix will hold them in place but not too deep so that they become submerged and provide no added skidding resistance. Rate of spread of the coated chippings is set so as to achieve full coverage. An upper value of $12.0 \, \text{kg/m}^2$ is used for 20 mm chippings, reducing to $9.5 \, \text{kg/m}^2$ for 14 mm nominal size chippings. Depth of embedment, or 'texture depth', is set at 1.5 mm. Post-compaction, this is measured using the sand patch test where a volume of sand (50 ml) is spread on the surface of the pavement in a circular patch of diameter, D, in millimetres, so that the surface depressions are filled with sand to the level of the peaks. The texture depth is obtained from the following formula:

$$TD \text{ (texture depth)} = 63600 \div D^2 \tag{7.2}$$

7.7   Materials in rigid pavements

7.7.1   General

A rigid pavement consists of a subgrade/subbase foundation covered by a slab constructed of pavement quality concrete. The concrete must be of sufficient depth so as to prevent the traffic load causing premature failure. Appropriate measures should also be taken to prevent damage due to other causes. The proportions within the concrete mix will determine both its strength and its resistance to climate changes and general wear. The required slab dimensions are of great importance and the design procedure involved in ascertaining them is detailed in Chapter 8. Joints in the concrete may be formed in order to aid the resistance to tensile and compressive forces set up in the slab due to shrinkage effects.

7.7.2   Concrete slab and joint details

As the strength of concrete develops with time, its 28-day value is taken for specification purposes, though its strength at 7 days is often used as an initial guideline of the mix's ultimate strength. Pavement quality concrete generally has a 28-day characteristic strength of $40 \, \text{N/mm}^2$, termed C40 concrete. Ordinary Portland cement (OPC) is commonly used. The cement content for C40 concrete should be a minimum of $320 \, \text{kg/m}^3$. Air content of up to 5% may be acceptable with a typical maximum water cement ratio of 0.5 for C40 concrete.

The effects of temperature are such that a continuous concrete slab is likely to fail prematurely due to induced internal stresses rather than from excessive traffic loading. If the slab is reinforced, the effect of these induced stresses can be lessened by the addition of further reinforcement that increases the slab's ability to withstand them. This slab type is termed continuous reinforced concrete (CRC). Alternatively, dividing the pavement into a series of slabs and providing movement joints between these can permit the release and dissipation of induced stresses. This slab type is termed jointed reinforced concrete (JRC). If the slab is jointed and not reinforced, the slab type is termed unreinforced concrete (URC). If joints are employed, their type and location are important factors.

Joints in concrete pavements

Joints are provided in a pavement slab in order to allow for movement caused by changes in moisture content and slab temperature. Transverse joints across the pavement at right angles to its centreline permit the release of shrinkage and temperature stresses. The greatest effect of these stresses is in the longitudinal direction. Longitudinal joints, on the other hand, deal with induced stresses most evident across the width of the pavement. There are four main types of transverse joints:

- Contraction joints
- Expansion joints
- Warping joints
- Construction joints.

Contraction occurs when water is lost or temperatures drop. Expansion occurs when water is absorbed or the temperature rises. The insertion of contraction and expansion joints permit movement to happen.

Contraction joints allow induced stresses to be released by permitting the adjacent slab to contract, thereby causing a reduction in tensile stresses within the slab. The joint, therefore, must open in order to permit this movement while at the same time prohibiting vertical movement between adjacent concrete slabs. Furthermore, water should not be allowed to penetrate into the foundation of the pavement. The joint reduces the thickness of the concrete slab, inducing a concentration of stress and subsequent cracking at the chosen appropriate location. The reduction in thickness is usually achieved by cutting a groove in the surface of the slab, causing a reduction in depth of approximately 30%. A dowel bar placed in the middle of the joint delivers the requisite vertical shear strength across it and provides load-transfer capabilities. It also keeps adjacent concrete surfaces level during temperature induced movements. In order to ensure full longitudinal movement, the bar is debonded on one side of the contraction joint.

Expansion joints differ in that a full discontinuity exists between the two sides, with a compressible filler material included to permit the adjacent concrete to expand. These can also function as contraction or warping joints.

Warping joints are required in plain unreinforced concrete slabs only. They permit small angular movements to occur between adjacent concrete slabs. Warping stresses are very likely to occur in long narrow slabs. They are required in unreinforced slabs only, as in reinforced slabs the warping is kept in check by the reinforcing bars. They are simply a sealed break or discontinuity in the concrete slab itself, with tie-bars used to restrict any widening and hold the sides together.

Construction is normally organised so that work on any given day ends at the location of an intended contraction or expansion joint. Where this proves not to be possible, a construction joint can be used. No relative movement is permitted across the joint.

The four transverse joints are shown diagrammatically in Figs 7.15 to 7.18. (It should be noted that, in all cases, reinforcement is required to support dowels/tie-bars during construction.)

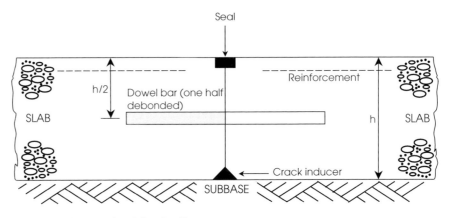

Figure 7.15 Contraction joint detail.

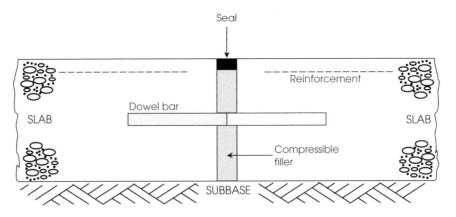

Figure 7.16 Expansion joint detail.

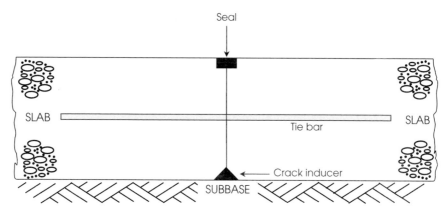

Figure 7.17 Warping joint detail.

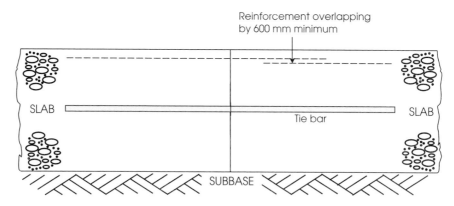

Figure 7.18 Construction joint detail.

Longitudinal joints may also be required to counteract the effects of warping along the length of the slab. They are broadly similar in layout to transverse warping joints.

7.7.3   Reinforcement

Reinforcement can be in the form of a prefabricated mesh or a bar-mat. The function of the reinforcement is to limit the extent of surface cracking in order to maintain the particle interlock within the aggregate.

In order to maximise its bond with the concrete within the slab, care must be taken to ensure that the steel is cleaned thoroughly before use. Because the purpose of the reinforcement is to minimise cracking, it should be placed near the upper surface of the pavement slab. A cover of approximately 60 mm is usually required, though this may be reduced slightly for thinner slabs. It is normally stopped approximately 125 mm from the edge of a slab, 100 mm from a longitudinal joint and 300 mm from any transverse joint.

Transverse lapping of reinforcement within a pavement slab will normally be in the order of 300 mm.

7.7.4   *Construction of concrete road surfacings*

There are a number of key issues that must be addressed in order to properly construct a concrete pavement. These include the positioning of the reinforcement in the concrete, the correct forming of both joints and slabs and the chosen method of construction, be it mechanised or manual.

Concrete paving is a dynamic and vigorous process, so it is imperative that the steel reinforcement is kept firmly in place throughout. In particular, chairs made from bent reinforcing bars permit the mesh or fabric reinforcement at the top of the slab to be secured throughout the concreting process. These chairs should be strong enough to take the weight of workers required to walk over them during the concreting process. Crack inducers must also be firmly connected to the subbase.

Dowel bars used in expansion and contraction joints are usually positioned on metal cradles so that they will not move from their required position while the concrete is being placed and compacted. These cradles, however, should not extend across the line of the joint. Tie-bars in warping joints are also normally part of a rigid construction that allows them to be firmly secured to the supporting subbase, while those in construction joints can be inserted into the side of the pavement slab and recompacted. Alternatively, both dowels and tie-bars can be vibrated into position.

Where the top of the pavement foundation consists of unbound material, it is possible that grout from the concreting process may leak into it. To prevent this occurring, and to minimise frictional forces, a heavy-duty polythene separation membrane is positioned between the foundation and the jointed concrete pavement.

The pavement slab can be constructed in one or two layers. Two layers could be employed where an air-entrained upper layer is being installed or for ease in the placement of reinforcement, where the reinforcement could be placed on the lower layer after it has been compacted, thereby obviating the need for supporting chairs.

On large pavement construction projects, continuous concreting is the most economic method of placement. Within it, the paver moves past joint positions, requiring that crack inducers, dowels and tie-bars be kept in position by methods referred to above. Purpose-built highway formwork is typically made of steel, held in position by road pins driven through flanges in the form and into the pavement foundation immediately below. Once the steel reinforcement and the formwork are in place, the concreting can commence.

Mechanised paving allows a higher quality concrete finish to be attained. The spreading, compacting and finishing of the pavement involves use of a fixed-form or slip-form paving train.

Fixed-form paving uses steel forms or a preconstructed concrete edge-beam to retain the concrete, using machine rails to support and guide the individual items of plant utilised in the pavement construction process. A train of machines, each individually operated, run along the rails, executing the basic tasks of:

- Spreading the concrete
- Compacting it by vibration
- Finishing the surface.

Machines for dowel and joint forming that leave the surface of the concrete with the required texture and the addition of curing compounds may also be included within the process. The machines themselves may be manually propelled, self-powered or towed along the rail.

Typical types of machinery used in a fixed-form paving train are:

(1) Feeder – receives concrete as it arrives at the required location
(2) Spreader – distributes the concrete across the full width of the pour in question, discharging it at a controlled rate
(3) Rotary strike-off paddles and compaction beams – regulate the concrete by trimming any irregularities in the concrete and vibrate its surface
(4) Dowel/tie-bar placers – place these elements in the appropriate joints either manually or by vibration
(5) Joint groove formers and finishers – grooves formed by a knife travelling within the plastic concrete (wet-formed). Otherwise, a vibrating blade can be used to form them when the concrete has hardened sufficiently
(6) Final finishing equipment – additional compaction and regulation of the concrete after the dowel and tie-bars have been put in place. (Machine uses two oblique finishing beams oscillating in opposite directions to achieve a uniform finish to the surface of the concrete)
(7) Curing compound sprayer
(8) Protective tentage.

The sequence of operations for two-layer placement with a fixed-form paver is illustrated in Fig. 7.19.

The previous method is analogous to manually placing the concrete. The process can also be completed without using fixed-forms. This process is called slip-form paving. It works on the basis that the sides of the pavement slab will support themselves before an initial set has been developed within the concrete. It produces a fully compacted slab. It cannot therefore be subsequently disturbed in order to place dowel or tie-bars, as the surrounding concrete could not then be properly made good. The slip-form paver spreads, compacts and finishes the concrete with only the forming and finishing of the joint grooves, texturing and curing done using other pieces of equipment.

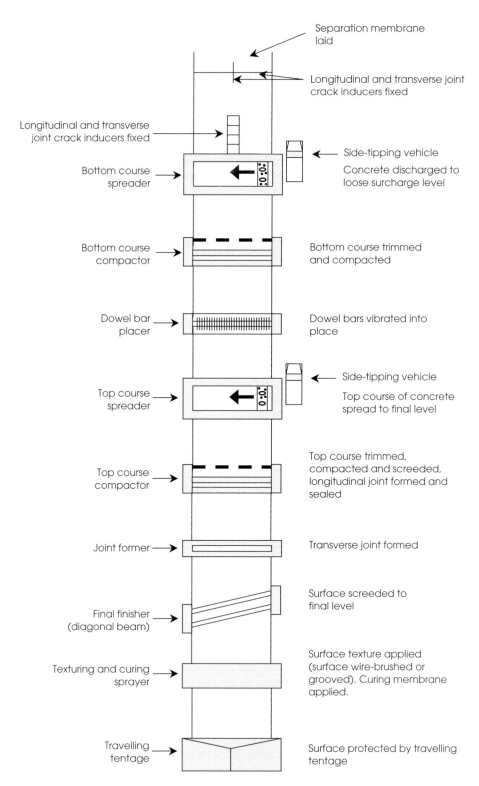

Figure 7.19 Diagrammatic representation of fixed-form paving train.

Both methods work on the basis that the full construction process is completed following one pass over the prepared foundation of the pavement. After the usual curing period the slab can then be subjected to normal traffic loadings.

Equally good results are possible with both types of paving machine. The slip-form paver has certain advantages/disadvantages associated with it:

- A higher output is achievable as less machinery is involved
- It will tend to be less expensive as labour costs will be lower due to the increased level of automation.

But:

- Edge slump may occur just after the concrete has left the paver
- Greater stockpiles of raw materials such as cement, steel mesh and aggregate are needed in advance of the operation in order to ensure continued output from the paving train
- The contractor operating it may be more vulnerable to weather conditions
- A minor quality control failure can cause the entire system to come to a sudden stop.

7.7.5   Curing and skid resistance

Concrete curing is an essential step in achieving a good quality finished product. It requires that both the temperature and moisture content of the mix be maintained so that it can continue to gain strength with time. If moisture is lost due to exposure to sunlight and wind, shrinkage cracks will develop. Such problems due to moisture loss can be avoided if the surface of the concrete is kept moist for at least seven days. This is usually achieved by mechanically spraying the finished surface, with exposure to rain avoided with the use of a travelling tentage as indicated in Fig. 7.19.

Immediately prior to the curing process, the surface should be textured in order to give it adequate wet-road skidding resistance. It is extremely important to get the texture to the correct level of quality at the time of construction as potential difficulties may arise with subsequent surface maintenance during its design life. Good skid resistance requires sufficient microtexture and macrotexture.

Macrotexture permits most of the rainwater caught between the tyres and the surface of the highway to drain rapidly and depends on grooves being developed on the surface of the mix in order to 'texture' it. Microtexture, on the other hand, depends on the use of fine aggregate within the mix. It must have abrasion-resistance properties such that the particles of sand stand proud of the matrix of the hardened cement paste while subject to traffic loading, therefore allowing it to penetrate the remaining film of water and maintain tyre contact with the surface. The required macrotexture of the surface is achieved by wire-brushing or grooving.

Wire-brushing is done either manually or mechanically from a skewed travelling bridge moving along the line of the pavement. The wire brush is usually a minimum of 450 mm long. Grooving is done using a vibrating plate moving across the width of the finished pavement slab forming random grooves. They have a nominal size of 6 mm by 6 mm, providing excellent surface water drainage properties. A high level of wet-road skid resistance is obtained by deep grooving, but problems may arise with higher tyre noise.

7.8   References

Black, W.P.M. & Lister, N.W. (1979) *The strength of clay fill subgrades: its prediction in relation to road performance.* Department of the Environment, Department of Transport. TRRL Report LR 889. Transport and Road Research Laboratory, Crowthorne, UK.

BSI (1984) BS 434 Parts 1 and 2 *Bitumen Road Emulsions (Anionic and Cationic).* British Standards Institution, London.

BSI (1990a) BS 1377 *Methods of test for soils for civil engineering purposes.* British Standards Institution, London.

BSI (1990b) BS 3690 Part 1 *Bitumens for Building and Civil Engineering: Specification for Bitumens for Roads and Other Paved Areas.* British Standards Institution, London.

BSI (1992) BS 594 *Hot rolled asphalt for roads and other paved areas.* British Standards Institution, London.

DoT (1994) Foundations, HD 25/94. *Design Manual for Roads and Bridges, Volume 7: Pavement Design and Maintenance.* The Stationery Office, London, UK.

DoT (1996) Traffic Assessment, HD 24/96. *Design Manual for Roads and Bridges, Volume 7: Pavement Design and Maintenance.* The Stationery Office, London, UK.

DoT (1999) General Information, HD 23/99. *Design Manual for Roads and Bridges, Volume 7: Pavement Design and Maintenance.* The Stationery Office, London, UK.

DoT (2001) Pavement design and construction, HD 26/01. *Design Manual for Roads and Bridges, Volume 7: Pavement Design and Maintenance.* The Stationery Office, London, UK.

O'Flaherty, C.A. (2002) *Highways: The location, design, construction and maintenance of pavements.* Butterworth Heinemann, Oxford.

Specification for Highway Works. London: HMSO, 1998.

Chapter 8

Structural Design of Pavement Thickness

8.1 Introduction

One of the basic requirements for a pavement is that it should be of sufficient thickness to spread the surface loading to a pressure intensity that the underlying subgrade is able to withstand, with the pavement itself sufficiently robust to deal with the stresses incident on it. Where required, the pavement should be sufficiently thick to prevent damage to a frost-susceptible subgrade. Thickness is thus a central factor in the pavement design process.

The thickness designs in this chapter are based on LR1132 (Powell *et al.*, 1984) for flexible pavements and RR87 (Mayhew & Harding, 1987) for rigid construction. However, these basic designs are modified and updated based on later research as detailed in HD 26/01 (DoT, 2001). The pavement thickness design methodologies for the two different categories of pavement type are treated separately below.

In order to reflect European harmonisation, the names of the various pavement layers have been altered within the context of thickness design, as seen in Fig. 8.1.

8.2 Flexible pavements

8.2.1 General

The pavement should be neither too thick nor too thin. If it is too thick, the cost will become excessive. If it is too thin, it will fail to protect the underlying unbound layers, causing rutting at formation level.

A flexible pavement is defined as one where the surface course, binder course and base materials are bitumen bound. Permitted materials include hot rolled asphalt (HRA), high density macadam (HDM), dense bitumen macadam (DBM) and dense bitumen macadam with 50-penetration bitumen (DBM50). Flexible composite pavements involve surface course and upper base materials bound with macadam built on a lower base of cement bound material (CBM). Wearing courses are either 45 mm or 50 mm of hot rolled asphalt or 50 mm of

Figure 8.1 Proposed renaming of pavement layers.

porous asphalt (PA). (If PA is used, it is assumed to contribute only 20 mm to the overall thickness of the pavement for design purposes.) The bitumen within dense bitumen macadam roadbases and basecourses must be at least 100 penetration grade, with hot-rolled asphalt containing 50 pen binder.

8.2.2   *Road Note 29*

Pavement thickness design methods have historically been empirically based, with the performance of pavements being analysed and design charts being compiled based on the information obtained from the on-site observations of researchers. This approach led to the publication of Road Note 29 (Department of the Environment, 1973). This document was based on scrutiny during the 1960s of the behaviour of sections of highway pavement along the A1 trunk road in Cambridgeshire trafficked by up to 10 million standard axles. It formed the basis for pavement design philosophy in the UK from then until the mid 1980s.

Road Note 29 (RN29) takes account of increasing axle loads and vehicle numbers while also differentiating between the performance characteristics of different roadbase materials. (The roadbase is assumed to satisfy the entire strength requirements for the entire pavement, with the surfacing considered to make no significant contribution to the strength of the pavement. The primary function of the surface material is to provide surface texture and regularity.) The RN29 procedure is best explained as a series of design steps.

Step 1
Determine the cumulative number of commercial vehicles expected to use the highway from its first day of use to the end of its design life, taken as 20 years.

Step 2
Determine the cumulative number of commercial vehicles expected to use the 'design lane' over its design lifetime. (The design lane is the most heavily trafficked lane in any given direction.)

Step 3
Determine the equivalent number of standard axles incident on the road over its design life, based on the commercial vehicle usage. Based on a standard axle

of 80 kN, the required value is obtained from the product of the cumulative number of commercial vehicles and a term called the damage factor which varies for different road types. The maximum value of this conversion factor is 1.08, used for motorways and trunk roads designed to cater for over 1000 commercial vehicles per day in each direction.

For a motorway:

$$\text{Equivalent No. of standard axles} = \text{No. of commercial vehicles} \times 1.08 \quad (8.1)$$

Step 4
Determine the subbase thickness. This is dependent on both the CBR of the subgrade and the cumulative number of standard axles over the design life of the highway. For a cumulative number of standard axles of 1 million (1 msa), a minimum subbase thickness of 150 mm is required where the CBR is greater than 6%, rising to 440 mm where the CBR is 2%. Where the CBR is less than 2%, an additional 150 mm of subbase should be added to that required for a CBR of 2%. The CBR of the subbase should be at least 30%. If the CBR of the subgrade is in excess of this value, no subbase need be used.

Step 5
Determine the roadbase and surfacing thickness. This parameter depends purely on the cumulative number of standard axles over the pavement's design life. For cumulative standard axles in excess of 10 million, the surfacing should be 100 mm thick (60 mm basecourse plus 40 mm wearing course). If dense bitumen macadam is used, a roadbase thickness of just under 150 mm is required to cater for 10 msa, giving a total bound thickness of 250 mm.

8.2.3 LR1132

Road Note 29 was the sole officially recognised pavement design methodology throughout the 1970s and early 1980s. While it was considered to be generally effective, it had certain inherent deficiencies:

- It was seen as unresponsive both to improvements in the quality of available raw materials and to changes in construction processes.
- The RN29 method is valid for designs up to 40 msa. Many highways were, by the mid 1980s, well in excess of 50 msa, with some approaching 150 msa over their 20-year design life.
- The 20-year design life implied that, after this period, a surface rut of 20 mm or more, or severe cracking or crazing had developed. The pavement was then considered to be in a failed state and in need of major strengthening or partial reconstruction. It has been shown that attempting to strengthen a pavement damaged to such an extent did not automatically result in satisfactory performance afterwards.

LR1132 (Powell *et al.*, 1984) revised RN29 by redefining pavement failure, thereby delivering a thicker but longer lasting highway likely to be in a less deteriorated state after 20 years.

The design criteria adopted by LR1132 were:

(1)   The subgrade must be able to sustain traffic loading without excessively deforming. This is achieved by limiting the vertical stress at formation level.
(2)   Bituminous or cement bound materials used in the flexible pavement must not be subject to fatigue cracking. This is achieved by limiting the horizontal tensile stresses at the bottom of the bituminous/cement bound roadbase.
(3)   The load spreading capability of granular subbases should be enough to provide an acceptable construction platform.
(4)   When a pavement is composed of a considerable depth of bituminous material, its creep must be restricted in order to stop the rutting which arises from internal deformation.

Some of the stresses referred to above are illustrated in Fig. 8.2.

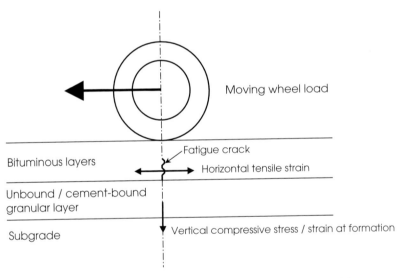

Figure 8.2 Critical stresses/strains in a bituminous highway pavement slab. (Crown copyright 1984)

In contrast with RN29, where the failure condition was presented as a 20 mm rut with severe cracking/crazing, LR1132 defined the end of a pavement's design life as indicated by a 10 mm rut depth or the beginning of cracking in the wheel paths. These less severe indications were chosen on the basis that they are the precursors of significant structural deterioration. They mark the latest time when the application of an overlay will have maximum effect and will be expected to make best use of the original structural quality of the pavement. In other words, the design life as thus defined is the latest time at which the application of an overlay will deliver another few years of high quality motoring. This is termed pre-emptive overlaying, a process carried out at the onset of critical structural conditions within the pavement. If application is postponed to a point later in the pavement's life, it may well have deteriorated to a stage where extensive pavement reconstruction will be required.

Since the LR1132 approach maximises the use of the existing pavement's strength, a pavement of more uniform strength will result. In addition, as deterioration can be predicted without too much difficulty, ultimate reconstruction can be more easily planned. This definition of design life results in LR1132 designing a pavement having an additional period of serviceable life before major reconstruction, a period that would not be available if Road Note 29 were used.

A design life of 20 years is normally employed.

Given the adoption of the design life concept as detailed within LR1132, the cumulative number of equivalent 80 kN standard axles to be carried during the design life of the highway must now be estimated. Observed or estimated 24-hour commercial vehicle flows must be converted to annual flows. If there is more than one lane in each direction, an allowance must be made for the proportion of this traffic travelling in the nearside lane, assumed to be the lane carrying the majority of commercial vehicles (RN29 makes this same assumption). The annual traffic is then multiplied by the vehicle damage factor – an estimator of the damage effect of an average commercial vehicle.

The design procedure can be summarised as follows.

Step 1
Calculate T_n, the total number of commercial vehicles using the slow lane over the n years design life, as follows:

$$T_n = 365 F_0 \frac{\left((1+r)^n - 1\right)}{r} P \qquad (8.2)$$

where
F_0 = initial daily flow (base year)
r = commercial vehicle growth rate
n = design life
P = proportion of commercial vehicles using the slow (nearside) lane
P = 1 if it is assumed that all vehicles use the nearside lane.

Step 2
Calculate the damage factor, D.
In order to convert T_n into equivalent standard axles, it must be multiplied by the vehicle damage factor, D, calculated for the mid year of the design life, F_m. The damage factor is calculated as follows:

$$D = \frac{0.35}{0.93^t + 0.082} - \left(\frac{0.26}{0.92^t + 0.082}\right)\left(\frac{1.0}{3.9^{(F \div 1550)}}\right) \tag{8.3}$$

where

F_m = number of commercial vehicles per day in one direction (mid-term year)
t = mid-term year minus 1945.

Step 3
Calculation of N, the cumulative number of standard axles

$$N = T_n \times D \tag{8.4}$$

Subgrade strength

The CBR test is taken as a direct measure of the strength of the in-situ subgrade material. Despite concerns regarding the limited accuracy of this test, it is utilised on the basis that it is widely used and accepted by both theorists and practitioners.

Subbase and capping layer

In terms of the overall structural strength of the pavement, the subbase is an extremely important layer. If the design life traffic volume is less than 2 msa, the CBR of the subbase must exceed 20%. If it is greater than 2 msa, the minimum CBR of the subbase rises to 30%. Use of a capping layer (with a lower specification than the subbase material) will allow a thinner layer of the more expensive subbase to be used in the pavement.

Table 8.1 indicates the thickness requirements for both subbase material alone and combinations of subbase and capping for different CBR values of the underlying subgrade material.

Table 8.1 Thickness of subbase and capping layers (Crown copyright 1984)

Layer	CBR of subgrade				
	<2	2	3	4	5+
Subbase thickness (mm)	615	400	310	260	225
Subbase + capping Comprising					
• Subbase thickness (mm)	150	150	150	150	225
• Capping layer thickness (mm)	600	350	350	350	—

Roadbase and surfacing

Figures 8.3 to 8.5 are illustrations of the required thicknesses for bituminous, wet-mix macadam and lean-mix concrete roadbases plus surfacings as detailed in LR1132.

Thickness of bound layers (mm)

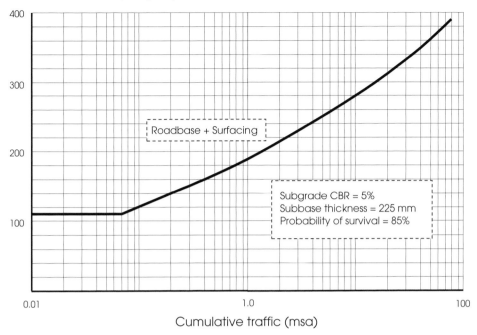

Figure 8.3 Design curve for highways with bituminous roadbase. (Crown copyright 1984)

Thickness of bound layers (mm)

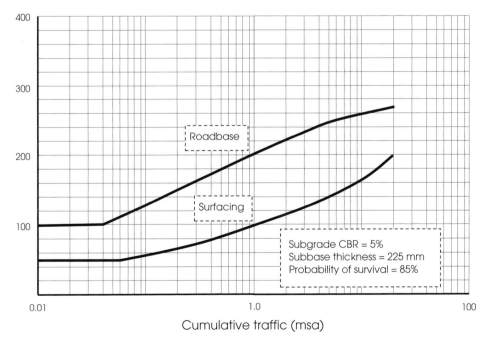

Figure 8.4 Design curve for highways with wet-mix roadbase. (Crown copyright 1984)

Thickness of bound layers (mm)

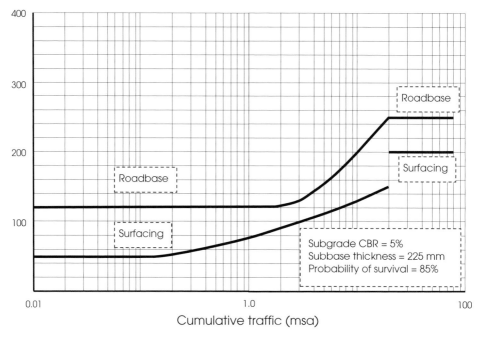

Figure 8.5 Design curve for highways with lean concrete roadbase. (Crown copyright 1984)

It should be noted from the design charts in LR1132 that not all highway pavement materials are suitable for traffic levels over 20 million standard axles. A standard long-life design is given for bituminous pavements with a lean-mix concrete roadbase. In the case of pavements with bituminous roadbases, designs for more than 80 msa are not included in the standard charts. At these traffic levels, the end result of deep-seated failure is generally severe traffic disruption. In order to avoid fatigue failure, a layer of rolled asphalt is placed at the position in the pavement where resistance to fatigue is most essential – at the bottom of the bound layer. This will help ensure that failure is likely to occur in the form of surface deformation rather than deep-seated fatigue failure. Consequent damage will be less extensive, confined to the upper regions of the highway pavement.

Figure 8.6 shows the general composition of a bituminous pavement designed to carry traffic in excess of 80 msa. The total thickness of the bound layer should be approximately 400 mm for cumulative traffic in the range of 100 msa.

Example 8.1 – Pavement design using LR1132

Design is required for a new pavement to be constructed in 2006. The annual average daily traffic for commercial vehicles on the opening day will be 1500.

Growth rate = 4%
Subgrade CBR = 3%
Design life = 20 years

Assume a bituminous roadbase is to be used.

Solution

$$T_n = 365 F_0 \frac{\left((1+r)^n - 1\right)}{r} P$$

where
  F_0 = 1500
  r = 0.04
  n = 20
  P = 1

Therefore:

$$T_n = 365 \times 1500 \frac{\left((1.04)^{20} - 1\right)}{0.04}$$

$$= 365 \times 1500 \times 29.78$$

$$= 16.3 \text{ million}$$

Mid-term year = 2016
Therefore:

$$t = 2016 - 1945$$

$$= 71$$

$$F = F_0(1+r)^{n/2}$$

$$= 1500 \times (1.04)^{10}$$

$$= 2220 \text{ commercial vehicles in the mid-term year}$$

$$D = \frac{0.35}{0.93^t + 0.082} - \left(\frac{0.26}{0.92^t + 0.082}\right)\left(\frac{1.0}{3.9^{(F \div 1500)}}\right)$$

$$= \frac{0.35}{0.93^{71} + 0.082} - \left(\frac{0.26}{0.92^{71} + 0.082}\right)\left(\frac{1.0}{3.9^{(2220 \div 1550)}}\right)$$

$$= 3.987 - (3.07 \times 0.14238)$$

$$= 3.987 - 0.4371$$

$$= 3.5499$$

Design traffic = 16.3 × 3.5499
             = 57.86 million standard axles say 60 msa

CBR = 3%

Therefore:

Subbase thickness = 310 mm subbase or 150 mm subbase plus 350 mm capping.

Roadbase + surfacing = 375 mm, assuming a bituminous roadbase (wet-mix roadbase is unsuitable and lean-mix roadbase results in surfacing of 200 mm on top of the lean-mix roadbase of 250 mm).

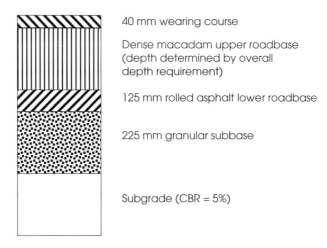

Figure 8.6 Pavement composition for traffic levels greater than 80 msa.

8.2.4   HD 26/01

The standard HD 26/01 (DoT, 2001) is based on LR1132 for flexible and flexible composite pavements, but with modifications and amendments to take account of more recent research, new materials and the observed functioning of in-service pavements.

The standard design life for all types of pavement is assumed to be 40 years, provided appropriate maintenance programmes are in place. For roads surfaced with asphalt, it is anticipated that surface treatment would be required every 10 years. (Note: in HD 26/01 all bituminous materials are covered by the generic term 'asphalt'.)

A 20-year design life may be seen as appropriate for less heavily trafficked schemes.

LR1132 was based on observations of highways over a 20-year period. Later research (Munn *et al.*, 1997) indicated that cracking and deformation is more likely to occur in the surfacing than deeper in the structure. Therefore, it was surmised that a well-constructed flexible pavement would have a very long structural life provided such signs of distress are treated before they affect the structural integrity of the highway. HD 26/01 notes that, for long-life flexible pavements designed to carry traffic for at least 40 years, it is not necessary to increase the thickness of the pavement beyond that required for 80 million standard axles.

For flexible composite pavements, the thickness of the cemented lower base can be reduced as the strength of the CBM is increased.

For fully flexible pavements, two design charts apply, one for recipe mixes, one for design mixes. Figure 8.7 is an illustration of the design thicknesses for four different types of base (roadbase) material:

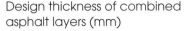

Design thickness of combined
asphalt layers (mm)

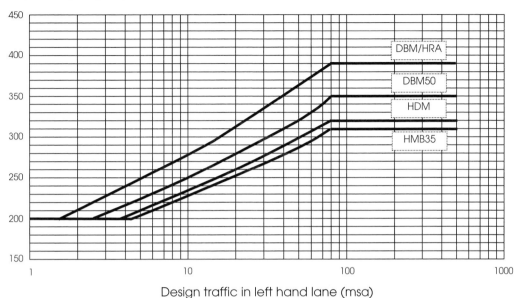

Design traffic in left hand lane (msa)

Figure 8.7 Design thickness for flexible pavements – recipe-based specification for roadbase (HD 26/01) (DoT, 2001).

- Dense bitumen macadam/hot rolled asphalt (DBM/HRA)
- Dense bitumen macadam with 50 penetration grade bitumen (DBM50)
- Heavy duty macadam (HDM)
- High modulus base with 35 penetration grade bitumen (HMB35).

The total thickness of a fully flexible pavement depends on the type of road-base used. A DBM/HRA base is the least stiff and therefore requires the greatest pavement thickness. As the stiffness progressively increases from DBM50 up to HMB35, a reduced pavement thickness provides an equivalent level of structural strength and integrity.

Similarly for flexible composite materials, the thickness of the pavement decreases as the strength of the CBM material used becomes greater. The design thicknesses for the asphalt (bituminous) layers within a flexible composite pavement are given in Fig. 8.8. The design thickness of the lower base varies from 150 mm to 250 mm depending mainly on the type of cement bound macadam used and the incident design traffic. A simplified diagram of the required thicknesses for the lower base of a flexible composite pavement is given in Fig. 8.8.

Design thickness
(mm)

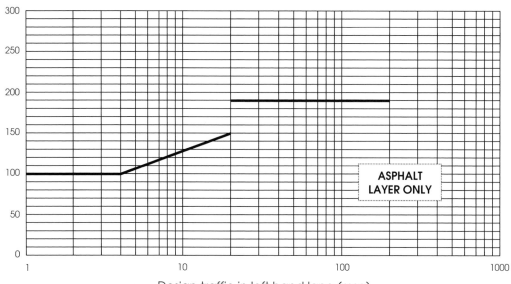

Design traffic in left hand lane (msa)

Design thickness
(mm)

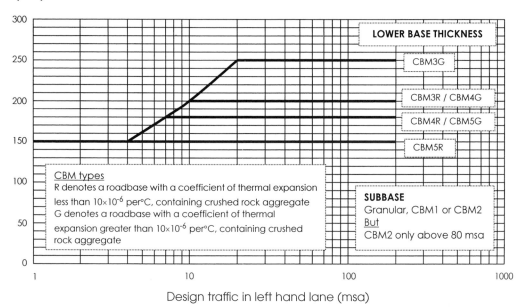

Design traffic in left hand lane (msa)

Figure 8.8 Design thicknesses for flexible composite pavements.

Example 8.2

A highway is envisaged to carry a traffic loading of 40 msa over its design life. Interpret the necessary asphalt thickness for a wholly bituminous pavement using the relevant chart from HD 26/01 (DoT, 2001).

Solution

From Fig. 8.7, the following required thicknesses can be deduced:

(1)   For DBM/HRA base: 350 mm
(2)   For DBM50 base: 310 mm
(3)   For HDM base: 290 mm
(4)   For HMB35 base: 280 mm.

Note: all thicknesses are rounded up to the nearest 10 mm in accordance with the guidance from HD 26/01.

Example 8.3

A highway is envisaged to carry a traffic loading of 30 msa over its design life. Interpret the necessary surfacing and lower base thicknesses for a flexible composite pavement using the relevant chart from HD 26/01.

Solution

From Fig. 8.8, the following information can be deduced.
   The total asphalt layer required is 190 mm. Typically, this would be composed of:

● 40 mm HRA wearing course (surface course), overlaying
● 50 mm basecourse (binder course), overlaying
● 100 mm roadbase (base).

For DBM/HRA base, required thickness is 350 mm
   The lower base layer required on a granular subbase is:

● 250 mm of CBM3G base, *or*
● 200 mm of CBM3R or CBM4G, *or*
● 180 mm of CBM4R or CBM5G, *or*
● 150 mm of CBM5R.

Note: CBM 5, 4 and 3 are the highest quality cement bound materials, usually prepared at a central plant from batched amounts of processed crushed gravel or rock.
   A very similar result is obtained from LR1132. If one examines Fig. 8.5 detailing required thicknesses for a concrete roadbase plus asphalt surfacing:
   For a cumulative traffic figure of 30 msa:

● Asphalt surfacing thickness = 200 mm
● Lean-mix concrete roadbase = 250 mm.

which is within 10 mm of the overall thickness recommended by HD 26/01.

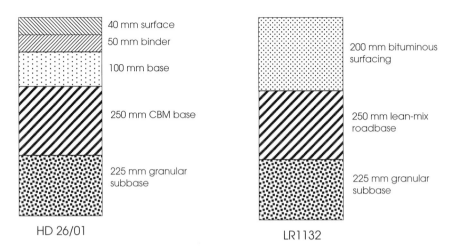

Figure 8.9 Comparison of HD 26/01 and LR1132 designs from Example 8.3.

8.3   Rigid pavements

8.3.1   *Jointed concrete pavements (URC and JRC)*

The Transport Research Laboratory publication RR87 (Mayhew & Harding, 1987) is, like LR1132, empirically based, taking account of full-scale concrete road experiments. The document concentrates on performance data for jointed unreinforced concrete (URC) and jointed reinforced concrete (JRC) slabs. The sites evaluated had service lives of up to 30 years and were subject to traffic loadings of over 30 million standard axles.

Mayhew and Harding derived two equations, each estimating pavement life based on a set of independent variables.

For URC pavements:

$$\text{Ln}(L) = 5.094\,\text{Ln}(H) + 3.466\,\text{Ln}(S) + 0.4836\,\text{Ln}(M)$$
$$+ 0.08718\,\text{Ln}(F) - 40.78 \tag{8.5}$$

where
Ln is the natural logarithm
L is the cumulative traffic carried by the highway during its design life (msa)
H is the slab thickness in mm
S is the 28-day cube compressive strength of the pavement concrete in MPa
M is the equivalent modulus of a uniform foundation providing identical slab support to the actual foundation, measured in MPa
F is the percentage of failed slabs

Definitions of failure in the slab include:

- A medium crack traversing the slab either transversely or longitudinally
- Medium width transverse/longitudinal intersecting cracks originating at a slab edge.

For reinforced concrete pavements, the relevant formula is:

$$\text{Ln}(L) = 4.786\,\text{Ln}(H) + 3.171\,\text{Ln}(S) + 0.3255\,\text{Ln}(M)$$
$$+ 1.418\,\text{Ln}(R) - 45.15 \qquad (8.6)$$

R is the quantity of high yield reinforcement within the reinforced concrete pavement slab, in millimetres per metre run. Normally, British Standard mesh is used, with R having the following three standard sizes:

- $R = 385\,\text{mm}^2/\text{m}$
- $R = 503\,\text{mm}^2/\text{m}$
- $R = 636\,\text{mm}^2/\text{m}$.

Estimates of the equivalent foundation modulus derived by Mayhew and Harding are detailed in Table 8.2 for different subbase materials and subgrade CBRs.

The slab thicknesses derived from Equations 8.5 and 8.6 are for concrete pavements that, under fully trafficked conditions, have a 50% probability of achieving the service life specified. A highway engineer may require a design incorporating a higher probability of survival; 85% is considered generally acceptable.

In order to increase the probability of survival, the slab thickness derived in either of the above equations should be increased by the value indicated by the graph in Fig. 8.10.

Table 8.2 Typical equivalent moduli for foundation (Crown copyright 1987)

Subbase						Subgrade		
Upper layer			Lower layer					
Type	Depth (mm)	Mod (MPa)	Type	Depth (mm)	Mod (mm)	CBR (%)	Mod (MPa)	Equiv. found mod
Granular			Capping	600	70	1.5	23	68
Type 1	150	150		350	70	2	27	65
	225	150	None			5	50	89
Lean				600	70	1.5	23	261
Concrete			Capping	350	70	2	27	268
(C10)	150	28 000		150	70	5	50	358
			None			15	100	683
Lean				600	70	1.5	23	277
Concrete			Capping	350	70	2	27	285
(C15)	150	35 000		150	70	5	50	383
			None			15	100	732

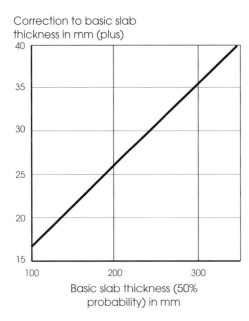

Correction to basic slab
thickness in mm (plus)

Basic slab thickness (50%
probability) in mm

Figure 8.10 Positive correction to slab thickness to increase probability of survival from 50% to 85%. (Crown copyright 1987)

In certain situations the pavement may not be fully trafficked as a margin of approximately 1 m on each side of the highway may be included to enable vehicles to pull over and stop. Such a margin, termed a tied shoulder, may also be used as a cycle track. If this section of the road is built as part of the overall slab, it will act to increase the overall effective strength of the concrete pavement, for practical purposes increasing its design life. On the other hand, if the designer has decided on a set life for the pavement, the thickness of the pavement can be decreased if the tied shoulders are allowed for.

Figure 8.11 gives the amount by which the slab thickness calculated in Equations 8.5/8.6 should be lessened if tied shoulders are to be taken into account within the overall design.

Figures 8.12 and 8.13 give final slab thickness values for jointed concrete pavements using the baseline Equations 8.5 and 8.6 along with the probability and tied shoulder allowances. The following specific assumptions have been made:

- The probability of survival for the slab over its full design life is set at 85%
- The proportion of unreinforced slabs failing, F, is set at 30%
- Two different reinforcement percentages, R, are referred to:
  - zero steel (unreinforced)
  - 636 mm^2/m
- Two equivalent foundation moduli, M, equal to 68 and 680 MPa, are assumed. (The family of curves for other values of equivalent foundation modulus can be found in Mayhew and Harding (1987).)
- The 28-day compressive strength, S, is set at a mean value of 48.2 MPa, the target value for the C40 mix (characteristic strength value of 40 N/mm^2).

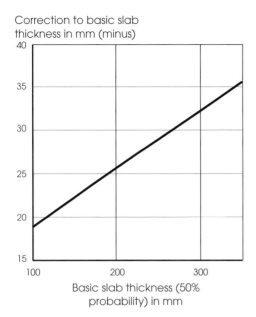

Correction to basic slab
thickness in mm (minus)

Basic slab thickness (50%
probability) in mm

Figure 8.11 Negative
correction to slab
thickness to make
allowance for
existence of tied
shoulders. (Crown
copyright 1987)

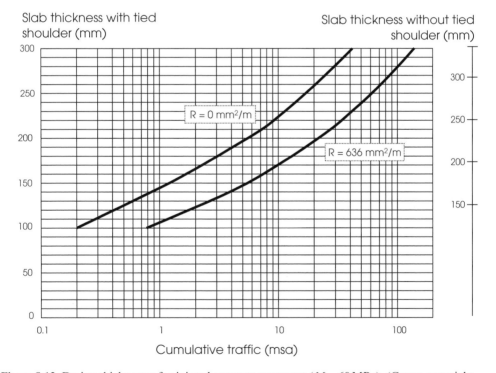

Figure 8.12 Design thicknesses for jointed concrete pavement ($M = 68$ MPa). (Crown copyright
1987)

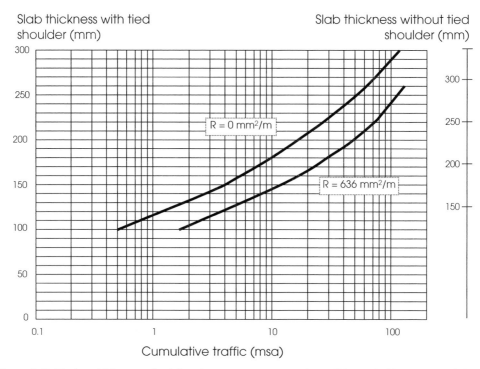

Figure 8.13 Design thicknesses for jointed concrete pavement ($M = 680$ MPa). (Crown copyright 1987)

HD 26/01 updated the design for jointed rigid pavements originally detailed in RR87. The design thicknesses for different traffic loadings are detailed in Fig. 8.14. The values derived assume the presence of a 1 m edge strip or tied shoulder adjacent to the most heavily trafficked lane. If this assumption cannot be made and the highway does not have a 1 m tied shoulder, a thicker slab will be required and the value derived from Fig. 8.14 must be increased by the amount given in Fig. 8.15.

Within a jointed reinforced concrete pavement, the quantity of reinforcement directly determines the joint spacing. If the slab is unreinforced and less than 230 mm thick, contraction joints should occur every 5 m. If the unreinforced slab is above 230 mm, these spacings reduce to 4 m.

If the jointed slab is reinforced, the maximum transverse joint spacing shall generally be 25 m. If limestone coarse aggregate is used throughout the slab, transverse joint spacing may be increased by one-fifth (HD 26/01).

It should be noted that, within the UK, rigid concrete construction is not generally recommended on trunk roads as it would require asphalt surfacing which, in the case of a jointed concrete pavement, has been observed to result in cracking and the need for subsequent maintenance. There are instances, however,

Example 8.4

A jointed reinforced concrete (JRC) pavement is to carry a traffic loading of 200 msa over its design life.

(1) Estimate the required slab thickness using the relevant chart from HD 26/01 assuming the use of tied shoulders
(2) Estimate the extra slab thickness required if tied shoulders are not used.

The reinforcement is set at 500 mm²/m.

Solution

From Fig. 8.14, the design thickness (assuming use of tied shoulders) is 280 mm.
From Fig. 8.15, if tied shoulders are not used, an additional 35 mm will be required.

Design thickness
(mm)

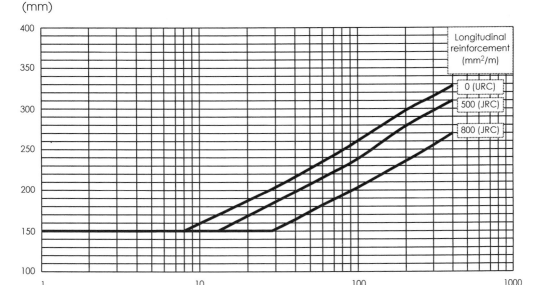

Design traffic in left hand lane (msa)

Figure 8.14 Design thickness for rigid jointed pavements (HD 26/01) (DoT, 2001).

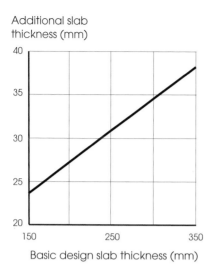

Additional slab thickness (mm)

Basic design slab thickness (mm)

Figure 8.15
Additional concrete slab thickness if 1 m tied shoulder not available (HD 26/01) (DoT, 2001).

where thin wearing course systems (TWCS) are acceptable. These have noise reducing properties and lend themselves to the application of further asphalt overlays during the pavement's life.

8.3.2   Continuously reinforced concrete pavements (CRCP)

A continuously reinforced concrete pavement is one in which the reinforcement is at a sufficiently high level that the slab can withstand the incident stresses without requiring the use of transverse joints. This slab type will require less maintenance than a jointed pavement and will span more effectively over areas where the underlying soil is weak. A CRCP ensures that a uniformly high stress level is maintained within the slab.

Cracks will develop within continuously reinforced slabs. This, however, will not give rise to concerns provided they are neither too wide nor too close together. Cracks should be no closer than 1.5 m but no further apart than twice this distance, as this would result in the cracks being at a width that would require significant maintenance. There is a strong likelihood that cracks less than 1.5 m apart will result in punching failure.

In the past there has been no generally accepted methodology for the design of continuously reinforced concrete slabs. Typically, slabs were designed as jointed reinforced, with the proportion of reinforcement fixed at a typical level of 0.6%. The thickness could then be reduced to allow for the longer design life arising from use of the continuously reinforced slab.

Most recently, HD 26/01 details slab designs for both rigid and rigid composite pavements. 'Rigid' denotes continuously reinforced concrete pavements (CRCP) with either no surfacing or a thin wearing course system with a minimum depth of 30 mm. Where noise levels are high due to the intensity of traffic on the highway, surfacing materials can significantly lower noise emissions generated by vehicles. Thin wearing course systems can be significantly quieter than concrete. It should be noted that, in England, rigid concrete construction is not permitted unless it has an asphalt surfacing. 'Rigid composite' denotes continuously reinforced concrete roadbase (CRCR) with 100 mm of asphalt surfacing. A thin wearing course can also be used with a binder course making up the remaining part of the required 100 mm. If porous asphalt is used over CRCR, it should be 50 mm thick over 90 mm of binder course, or 50 mm thick over 60 mm of binder course with the concrete slab thickness increased by 10 mm.

In both types of continuous slab, the reinforcement should be 60 mm below the surface in order to limit the opening of cracks. Laps should be 50 times the bar's diameter. Longitudinal reinforcement in CRCP either without surfacing or with a 30 mm thin wearing course system (TWCS) shall be 0.6% of the cross-sectional area. The required diameter is 16 mm (deformed steel). Transverse reinforcement shall be 12 mm in diameter, spaced at 600 mm centres. For CRCR slabs, the longitudinal reinforcement shall comprise 0.4% of the cross-sectional area, composed of 12 mm diameter deformed bars.

Figure 8.16 details the design thicknesses for both continuously reinforced concrete pavements and continuously reinforced concrete roadbases.

It should be noted that these figures assume the existence of a 1 m edge strip. If this is not available, the slab should be increased using Fig. 8.15.

Example 8.5

A continuously reinforced concrete pavement is to carry a traffic loading of 200 msa over its design life.

Estimate the required slab using:

(1)   CRCP
(2)   CRCR.

Give some surfacing options.

Solution

(1)   Required reinforced concrete pavement thickness of 230 mm with 30 mm of thin wearing course system (TWCS)
(2)   Required reinforced concrete base thickness of 210 mm with the 100 mm of surfacing comprising either 45 mm hot rolled asphalt (HRA) surface course overlaying 55 mm of dense bitumen macadam (DBM) binder course *or* 15 mm of TWCS overlaying 85 mm of DBM binder course.

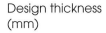

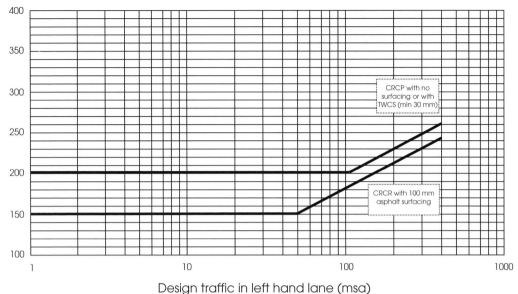

Figure 8.16 Design thickness for continuously reinforced concrete pavements and continuously reinforced concrete roadbase (HD 26/01) (DoT, 2001).

8.4   References

Department of the Environment (1973) *Road Note 29 – A guide to the structural design of pavements for new roads*, 3rd edn. The Stationery Office, London.

DoT (2001) Pavement design and construction. HD 26/01. *Design Manual for Roads and Bridges, Volume 7: Pavement Design and Maintenance*. The Stationery Office, London, UK.

Nunn, M.E., Brown, A., Weston, D. & Nicholls, J.C. (1997) *The design of long-life flexible pavements for heavily trafficked roads*. Department of the Environment, Department of Transport, TRRL Report 250. Transport and Road Research Laboratory, Crowthorne, UK.

Powell, W.D., Potter, J.F., Mayhew, H.C. & Nunn, M.E. (1984) *The structural design of bituminous roads*. Department of the Environment, Department of Transport, TRRL Laboratory Report LR1132. Transport and Road Research Laboratory, Crowthorne, UK.

Mayhew, H.C. & Harding, H.M. (1987) *Thickness design of concrete roads*. Department of the Environment, Department of Transport, TRRL Research Report RR87. Transport and Road Research Laboratory, Crowthorne, UK.

Chapter 9

Pavement Maintenance

9.1 Introduction

Highway pavements, once constructed, will not last forever. After a time, signs of wear will appear. These signs include cracking, rutting and polishing of the road's surface. A point will arrive where the wear and tear is at such an advanced stage that the integrity of the pavement and hence the standard of service provided by it has diminished. Maintenance is required at this point to prolong the highway's useful life. Loss of skidding resistance and loss of texture are forms of deterioration eventually suffered by all highway pavements.

In order to carry out the maintenance in as cost-effective a manner as possible, a logical coherent procedure must be adopted in order to select the most effective form that the maintenance should take, together with the optimum time at which this work should be undertaken. Minor maintenance may be sufficient to maintain the required standard of service for the motorist. However, in situations where major structural strengthening is required, a comprehensive structural investigation is vital in order to assist in the completion of the required detailed design (HD 30/99) (DoT, 1999).

9.2 Forms of maintenance

For bituminous roads, maintenance can be either minor or major. Minor maintenance takes the form of patching. It allows defective materials, particularly those in the surface courses of the pavement, to be replaced. If done properly, it can restore the stability and riding quality of the surface, arresting its deterioration and extending its serviceable life. It is an integral part of highway maintenance and makes sound economic sense.

Patching can remedy the following defects:

- Substandard drainage or some other problem related to the subgrade which will cause the failure of the pavement's foundation
- The aging of the bituminous surface, causing its break-up with the consequent formation of potholes and areas of crazing

- Decreased load bearing capacity of the pavement due to the ingress of water and damage due to frost.

Patching involves the repair of random areas of substandard pavement, not continuous widths/lengths.

Major maintenance of bituminous pavements may involve removing all or part of the surface using a planer and resurfacing the road, or laying a layer of bitumen over the existing one. The process of overlaying is dealt with in detail within this text as bituminous pavements are designed under LR1132 on the basis that, at the end of their design life, their structural integrity is such that the application of an overlay will significantly extend their useful life.

Both overlaying and resurfacing are carried out for the following common reasons (HD 31/94) (DoT, 1994c):

- To strengthen the highway pavement
- To replace defective materials
- To restore skidding resistance
- To improve riding quality.

Design of an overlay (minimum 50 mm thick) involves estimating the thickness required to deliver the required additional life to the pavement slab.

For concrete pavements, maintenance can take the following forms:

- Surface treatment – having assessed the skidding resistance of the pavement (using SCRIM – see section 9.7), it can be restored to an acceptable level by methods such as surface dressing consisting of epoxy resin-based binder and calcined bauxite chippings or mechanical roughening of the worn concrete surfacing by abrasive blasting or scabbling.
- Joint repairs – defective joint seals allow foreign matter to enter between the slabs and penetrate the lower levels of the pavement. Replacement of the seal will rectify this defect.
- Structural repair – cracks in the slab or movement at the joints can lead to outright failure if not treated. These are defined in terms of increasing severity as narrow (less than 0.5 mm), medium (between 0.5 mm and 1.5 mm) or wide (greater than 1.5 mm). Narrow cracks generally require no immediate action, but may require to be sealed. Medium cracks require a groove to be formed and a seal to be applied, while wide cracks may require a full-depth repair or bay replacement.
- Strengthening – may be required in order to extend the pavement's life due to increased traffic levels or because the slab in its present unimproved state is unable to carry the traffic predicted. This is accomplished using overlays. This is a more difficult process than for bituminous pavements due to the existence of joints. Therefore, concrete overlays are carried out either by forming joints in the overlay at the same location as those underneath or by separating the two concrete layers using a regulating layer of bituminous material. If a bitumen-based overlay is used, added thickness may be

required to deal not only with the structural requirements of the slab but also to counteract reflection cracking resulting from movements in the joints within the underlying concrete slab (HD 32/94) (DoT, 1994d).

9.3   Compiling information on the pavement's condition

An inventory of highway condition data is updated on a routinely regular basis in order to determine the condition of the pavement and whether its level of deterioration is such that remedial action is necessary. The data compiled will allow trends in the structural condition of a pavement to be established.

The four major types of routine assessment are (HD 30/99) (DoT, 1999):

- Visual condition surveys
- High speed road monitor
- Deflectograph
- SCRIM (see section 9.7).

The first three are dealt with here, and the fourth later in the chapter.

Visual condition surveys

These record defects that remain undetected by machine-based methods. As the method is slow and laborious, it is usually targeted on specific areas in particularly poor condition. The surveys provide factual information for deciding on the most appropriate structural treatments, and identify sections of highway suitable for remedial treatment. Planning for long-term treatment can thus be undertaken, with performance of the pavements being monitored and priorities for treatment being established on the basis of the database compiled.

In past times a Marshall Survey represented the usual method for preparing an inventory of the condition of sections of highway. It involved groups of technical staff compiling information throughout their area on the condition of carriageways, kerbs, etc. Nowadays, the information can be collated using the computer program CHART (Computerised Highway Assessment of Ratings and Treatments), a management system for assisting maintenance engineers. It provides information on the lengths of highway that are substandard, the treatments required and the relative priorities for treatment.

For every 100-metre section of highway, CHART convert defects information into numerical ratings for each existing defect. Those defects requiring treatment can then be identified. The ratings, which are subjectively based, indicate the relative urgency of the need for treatment.

Typical output is:

Map CHART – the quantity of defects are shown by chainage both along and across the highway

Section CHART – a summary indicating the average and maximum ratings

Treatment length CHART – only shows those sections where treatment has been
   recommended (separate lists are compiled for each form of treatment)

Subsection rating and treatment CHART – lists the sections recommended for
   treatment. Main recommendations for treatment are given.

A computerised system such as CHART has not yet been developed for rigid
pavements. Nonetheless, the aim of the procedure in this instance must be to
accurately record all relevant features observed, be that the condition of the carriageway itself or related matters such as drainage or earthworks problems.

HD 29/94 (DoT, 1994a) provides guidance on how the information from
visual surveys should be interpreted. A summary of these interpretations is given
in Tables 9.1 and 9.2.

Table 9.1 Interpretation of visual data for bituminous surfaces

Observations	Interpretation
Single longitudinal cracks in wheelpath	Indicate the onset of structural failure in pavements greater than 200 mm thick. These do not heal and deterioration will be progressive
Multiple wheelpath cracking and crazing	Narrow cracks imply condition is not near failure. Wider cracks imply advanced failure of a thicker structure
Longitudinal cracking outside the wheelpath	In all likelihood the location of a construction joint
Short transverse cracks	Unlikely to be structurally significant. Has probably started at the surface and will progress slowly
Long transverse cracks	Indicate a discontinuity in a lower layer, possibly a construction joint in a bituminous material

Table 9.2 Interpretation of visual data for concrete pavements

Observations	Interpretation
Longitudinal cracks	Indicate either the onset of structural failure, differential settlement or compression at the joints. Deterioration is likely to be rapid. Sealing cracks can reduce the rate of propagation
Mid-bay/third bay cracks	These are thermally induced in URC pavements, a consequence of a joint malfunctioning. They can result in serious distress if left unsealed
Joint damage	Sealant damage and spalling indicate excessive movement at the joint. Foundation damage and voiding may have occurred

It should be noted that narrow (non-structural) ruts indicate that the wearing
course is probably deforming under traffic. No effect on the lower layers will
result. With wide (structural) ruts, however, deformation is taking place deep
within the pavement and structural damage is occurring. It is possibly the result
of too much moisture in the unbound granular layers and excessive stress incident on the subgrade.

High speed road monitor

The high speed road monitor (HRM) measures parameters such as riding quality, texture and rutting. It consists of a van and 4.5 metre-long trailer fitted with four laser sensors along the nearside wheelpath. It allows the condition of the highway's surface to be assessed under conditions of normal traffic with speeds up to 95 km/h.

HRM is not designed to replace other slower visual techniques, but will allow them to be more effectively used by targeting them on the more vulnerable sections of pavement as identified by this methodology.

The four lasers within the HRM provide data on the longitudinal profile and macrotexture of the pavement. In addition, a laser mounted at the mid-point of the trailer axle is used along with the trailer wheels to measure the average rut depth in the nearside and offside wheelpaths. Inclinometers mounted on the trailer estimate the crossfall and gradient on the highway. Horizontal radius of curvature can also be estimated. A device fitted to the nearside trailer wheel provides distance information to which all HRM measurements are referenced.

The following major parameters are measured within HRM:

- Longitudinal profile – profile unevenness is used to assess riding quality for the road user. It measures deviations away from a moving datum which represents an average profile along the highway
- Rutting – HRM records the average rut depth to the nearside and offside wheelpaths
- Macrotexture – this is the coarser element of the surfacing, formed either by the aggregate particles in the surfacing or grooves in the concrete surface. It contributes to skidding resistance, particularly at high speeds, mainly by providing drainage paths for the water
- Alignment parameters – gradient, crossfall and radius of curvature measurements are estimated to an accuracy of ±10%.

HRM surveys are only beginning to be used on a routine basis. Results from them should be used for general guidance only. Its main function is to identify specific discrete locations where more established but more costly and time consuming methods of analysis should be used.

Deflection beam/deflectograph

The deflection beam is a widely accepted instrument for assessing the structural condition of flexible highway surfacings. Originated by Dr A.C. Benkleman and developed by the Transport Research Laboratory (Kennedy *et al.*, 1978a), it involves applying a load to the pavement's surface and monitoring its consequent vertical deflection. As a loaded wheel passes over the pavement surface, the deflection of the slab is measured by the rotation of a long pivoted beam touching the surface at the point where the deflection is to be determined. The deflection that occurs at the time and position of application of the load is termed the *maximum deflection*. The deflection that remains after the load is removed (permanent deflection) is termed the *recovery deflection*. It is the cumulative effect of the latter type of deflection that leads to cracking, rutting and ultimately failure of a pavement.

The rear axle of a dual-axle lorry is loaded symmetrically to 6250 kg. It has two standard closely spaced twin wheel assemblies at its rear, with approximately 45 mm between their walls in each case. The test commences with the loaded wheels moved into the starting position (see Fig. 9.1). An initial reading is then taken. The measurement cycle consists of the vehicle moving forward past the front end or tip of the beam to a point where the wheels are a minimum of 3 m past the end of the tip. During this time the deflection of the beam is monitored and the maximum value recorded. The deflection remaining at the end of the test is the recovery deflection for that point. The readings are temperature corrected. Half the sum of the maximum deflection plus the recovery deflection yields the deflection value, expressed in hundredths of millimetres.

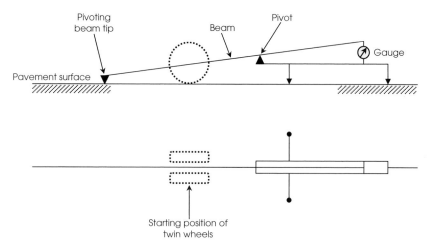

Figure 9.1 Measurement of pavement deflection using the Benkleman beam.

The deflection beam is used on short lengths of highway of 1 km or less.

The deflectograph provides a quicker method of assessment and is more suitable for the assessment of long sections of highway. It operates on the same principle as the deflection beam, but the measurements are made using transducers instead of dial gauges and all measurements are made in pairs, one in each of the wheel tracks. Loading is provided by two beams mounted on an assembly attached to the vehicle. An operating cable connects the deflection beam system to the lorry (Position A in Fig. 9.2). As the vehicle moves, the cable is let out at such a rate that the beam assembly remains stationary on the highway allowing the deflection under the rear wheels to be estimated (Position

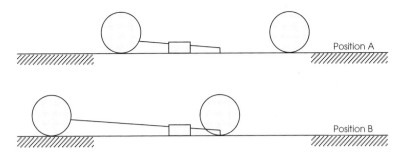

Figure 9.2 Operation of deflectograph.

B in Fig. 9.2). The vehicle then draws the beam assembly forward to the next measurement position where the cycle is repeated.

The correlation between the results from the two mechanisms is given in Fig. 9.3. The deflectograph and the Benkleman beam give similar but not identical results, with the deflectograph giving lower readings. The results are reasonably independent of temperature changes, within a range of 10°C to 30°C. The relationship illustrated is derived from measurements of pavements with rolled asphalt, bitumen macadam and tarmacadam surfacings, bitumen-bound, cement-bound and crushed stone roadbases and subgrades with CBR values between 2.5 and 15%.

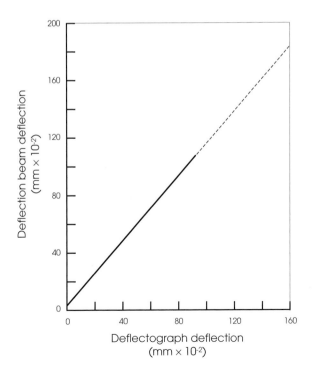

Figure 9.3 Correlation between deflectograph and deflection beam measurements. (Crown copyright 1978)

9.4   Deflection versus pavement condition

There is an established relationship between pavement deflection and pavement condition as expressed by repetitive loading of it by vehicular traffic. As the cumulative traffic increases with resulting damage to the pavement, higher deflections will be observed as its structural condition deteriorates. Deflection remains generally stable until the 'investigatory condition' is reached where deflections will start to increase with applied loading and cracks will begin to appear. The 'failed condition' is arrived at when rapid increases in deflection are experienced.

In the UK, deflection histories for highways have been derived from the monitoring of deflection for flexible pavements. Research Report 833 (Kennedy *et al.*, 1978a) developed performance charts relating deflection to cumulative traffic for three different types of roadbase: granular, cement bound and bituminous. The relationships had been found to be valid for subgrade CBR ranging from 2.5 to 15. Each illustrates deflection trends from its construction to the onset of structurally critical conditions, referred to above as the investigatory phase, and contains four envelope curves relating to different probabilities that the critical life of the pavement will be attained (the point at which strengthening will be required to prolong its useful life).

Critical conditions are defined by TRRL Laboratory Report 833 as cracking confined to a single crack or extending over less than 50% of the wheel track and/or rutting 19 mm or less. By contrast, failed conditions are defined as interconnected multiple cracking extending over the greater part of the width of the wheel path and/or rutting 20 mm of more. Effective maintenance requires that remedial work is commenced before the onset of critical conditions. Such an approach will deliver greater value for money than delaying work until the failed condition is reached and total reconstruction becomes the only way of prolonging the highway's useful life.

Figure 9.4 illustrates the deflection/cumulative traffic relationship for bituminous roadbases as given in TRRL Laboratory Report 833.

(Note: standard deflection relates to the equivalent deflection beam deflection at a temperature of 20°C. If the deflection measurement is made with a deflectograph it will be necessary to correct this value to a deflection equivalent to that obtaining at a pavement temperature of 20°C (LR833)).

Example 9.1

A pavement has a current rate of flow of 950 commercial vehicles per day using the slow lane. It is 10 years old. The damage factor over the time since opening is set at 0.9. The historic growth rate was estimated at 3%.

The standard deflection was measured as 50×10^{-2} mm.

Assume a bituminous roadbase and surfacing:

Contd

Example 9.1 Contd

(1)   Estimate the life in millions of standard axles that the pavement has a 0.5 probability of achieving
(2)   Estimate the life in millions of standard axles that the pavement has a 0.9 probability of achieving.

Solution

If the daily flow of commercial vehicles after 10 years is 950 vehicles, then, assuming a growth rate of 3%, the initial flow on the opening day is estimated using the following formula:

$$F_0 = F_n/(1 + r)^n$$

where
$\quad F_0$ = initial flow on opening day
$\quad F_n$ = flow after n years
$\quad r$ = traffic growth rate

Substituting the above values into Equation 9.1, the following initial flow is obtained:

$$F_0 = 950/(1.03)^{10}$$
$$= 706.89 \text{ commercial vehicles}$$

Addis and Robinson (1983) have shown that the total number of commercial vehicles T_n, using the slow lane over a given time span n can be expressed in terms of the initial daily flow F_0, the growth rate r as follows:

$$T_n = 365F_0/((1 + r)^n - 1)/r)$$

Therefore, since:

$\quad F_0 = 706.89$
$\quad r = 0.03$
$\quad n = 10$

$$T_n = (365 \times 706.89)/((1.03)^{10} - 1)/0.03)$$
$$= 2.96 \text{ million commercial vehicles}$$

This figure multiplied by the historic vehicle damage factor gives the total number of standard axles incident on the pavement since its opening:

Cumulative standard axles = $2.96 \times 0.9 = 2.66$ million standard axles (msa)

From Fig. 9.4, it can be seen that a pavement with a bituminous roadbase and a standard deflection of 50×10^{-2} mm has a 0.5 probability of achieving

Contd

Example 9.1 Contd

a life of approximately 3.8 msa. If the probability is increased to 0.9, the life reduces to just under 3.0 msa.

Therefore, the pavement is approximately 70% towards the design life it has 0.5 probability of attaining and approximately 90% towards the design life it has 0.9 probability of achieving.

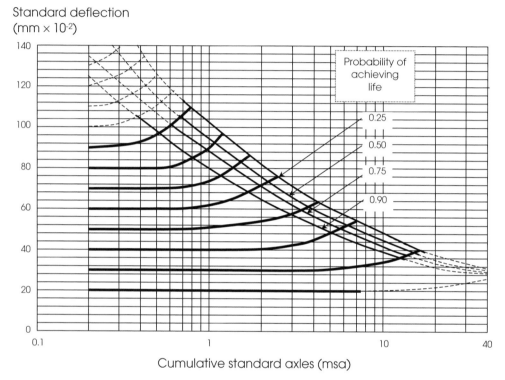

Figure 9.4 Relationship between standard deflection and life for pavements with a bituminous roadbase (source: LR833). (Crown copyright 1978)

9.5   Overlay design for bituminous roads

The deflection incident on a flexible pavement can be lessened and its structural strength enhanced by adding a bituminous overlay. The reduction in deflection is influenced by three factors:

- The thickness of the overlay
- The deflection of the existing pavement
- The elastic properties of the overlay material.

The first two factors in particular have been found to be the major determinants of overlay thickness. Subgrade strength and the original thickness of the pavement were not found to be major influential factors.

Within this text, an empirical overlay design method for flexible pavements as detailed in LR833 is presented. Other methods such as component analysis (Asphalt Institute, 1983) and analytically based processes (Claessen & Ditmarsch, 1977) are also available but are not addressed here.

The information in LR833 is based on the results of full-scale highway experiments where the surface has been overlaid with rolled asphalt at a time when, some sections of the road were approaching failure.

The LR833 method is based on assigning a representative deflection to a section of highway – termed the 85th percentile value (85% of the deflections found are less than this value). This value is then used in association with the family of curves as given in Fig. 9.4 to deliver a predicted life. If this is less than required, a bituminous overlay can be considered. Any residual life derived will be the period of time remaining before the application of an overlay or some other form of remedial action will be necessary.

The overlay design procedure is contained within a number of charts, each specifying the thickness of rolled asphalt overlay required to strengthen a pavement of given deflection so that the extended life required by the designer can be achieved. Charts for four different roadbases are contained in LR833:

- Granular roadbases whose aggregates have a natural cementing action
- Non-cementing granular roadbases
- Cement-bound roadbases
- Bituminous roadbases.

For a given roadbase material, different overlay thicknesses will be derived depending on the probability assigned to the assessment of the remaining life of an existing pavement and to the design of the extension of its life by overlaying. Therefore, LR833 supplies design charts for two different levels of probability, 0.5 and 0.9, for each roadbase material.

An illustration of the design chart for bituminous roadbases with a 0.5 probability of achieving their desired life is given in Fig. 9.5.

Example 9.2 – Overlay design using LR833

Taking the pavement detailed in Example 9.1. If a further 15 years of service is required from the pavement and the vehicle damage factor is expected to be 1.05 over this time span, estimate the thickness of bituminous overlay required to provide sufficient strengthening.

The future growth rate for traffic is set at 4%. Assume a 0.5 probability level.

Solution

Assuming that the present flow is 950 vehicles per day, using Equation 9.2, the estimation of cumulative traffic over the next 15 years is:

$$T_n = 365F_0/((1 + r)^n - 1)/r)$$

Contd

Example 9.1 Contd

Therefore, since:

$$F_0 = 950$$
$$r = 0.04$$
$$n = 15$$
$$T_n = (365 \times 950)/((1.04)^{15} - 1)/0.04)$$
$$= 6.94 \text{ million commercial vehicles} \times 1.05$$
$$= 7.29 \text{ msa}$$

Therefore, total life for which strengthening is required is estimated as:

$$\text{Cumulative required life} = 2.66 + 7.29$$
$$= 9.95 \text{ msa}$$

From Fig. 9.5, a 60 mm overlay is required (value given to the nearest 10 mm). This figure is based on the use of hot rolled asphalt (HRA) within the overlay. If open textured macadam is used, the value derived from Fig. 9.5 must be multiplied by a factor of two. If dense bitumen macadam (DBM) with 200-penetration grade bitumen binder is used, this factor is reduced to 1.3. If DBM with 100-pen bitumen binder is employed, no factor is required (i.e. the same thickness as for HRA will suffice).

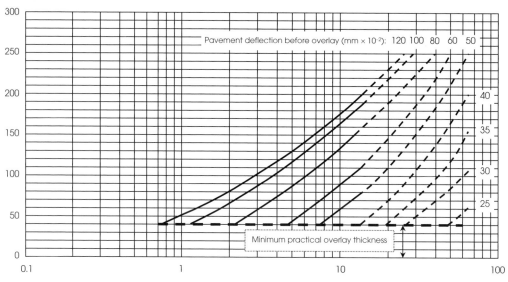

Figure 9.5 Overlay design chart for highway pavements with bituminous roadbases (probability = 0.5) (source LR833). (Crown copyright 1978)

9.6   Overlay design for concrete roads

The general guidance given in HD 30/99 regarding the overlaying of concrete pavements is given in Fig. 9.6. The existence of joints in the existing concrete pavement presents certain problems: reflection cracking in bituminous overlays (explained below) and, in the case of concrete overlays, the accommodation of differential movement across the existing joints. In the case of the latter, the solution lies either in the formation of the overlaying joints at the same location or the unbonding/debonding/separating of the overlay from the underlying existing pavement.

Whatever the overlaying material, bituminous or concrete, the underlying rigid pavement must give a secure, durable and homogeneous platform on which the overlay can be placed.

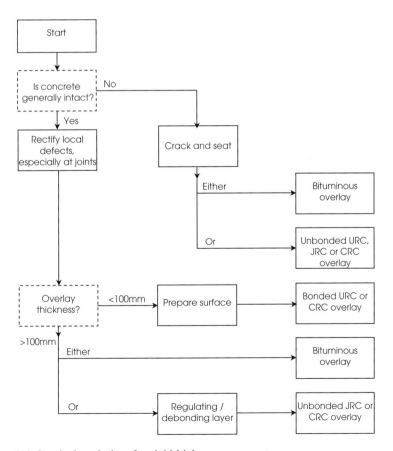

Figure 9.6 Overlaying choices for rigid highway pavements.

9.6.1   Bitumen-bound overlays placed over rigid pavements

For unreinforced and reinforced jointed pavements (URC and JRC), the overlay thickness may be controlled by the need to either reduce or delay the onset of reflection cracking rather than for basic structural reasons. Reflection cracking arises where the cracking pattern in the underlying pavement 'comes through' the overlay giving a similarly shaped pattern on the surface. A 180 mm thick overlay is usually required to counteract this condition.

If existing pavements have deteriorated markedly, it may be appropriate to 'crack and seal' before overlaying. This preliminary procedure involves inducing fine vertical transverse cracks in the existing concrete pavement, thereby reducing the load distributing properties of the slab but assisting in the controlling of reflection cracking.

9.6.2   Concrete overlays

Concrete overlays are not widely used within the UK (HD 30/99). This treatment results in the existing pavement having a longer life and improved surface characteristics as well as providing improved strength characteristics. A concrete overlay will function better if the existing pavement slab beneath it provides a good, firm foundation. Use of a thick concrete overlay is not suitable where the existing foundation is in a poor condition and where indications are that the subgrade is weak, in which case reconstruction may be the only viable option (HD 30/99). For a concrete overlay to be successfully used, it is imperative that the foundation is in good condition. Any voiding beneath a rigid slab should be filled with grout. In the case of a flexible/flexible composite pavement, any cracks should be repaired to ensure a good supporting structure.

Accurate information on the structural condition of the original pavement structure is required in order to make optimum use of the concrete overlay design method. The equivalent surface foundation modulus (ESFM) is the variable most often used to state, in quantitative terms, the structural integrity of the foundation.

Equivalent surface foundation modulus (ESFM)

This modulus was outlined in RR 87 (Mayhew & Harding, 1987). It evaluates the support offered by the foundation to a concrete pavement slab in terms of Young's modulus of an equivalent uniform elastic foundation of infinite depth. Equivalence is defined within RR 87 as 'the uniform elastic foundation with the same surface deflection, under a standard wheel load, as that of the actual foundation'. Equivalent thickness is thus used to transform the foundation into a corresponding single layer supported on a uniform elastic medium. Use of

Boussinesq's equations (1885) allows the surface deflection and hence the elastic modulus to be estimated.

The equivalent moduli for a number of foundations are given in Table 8.2. It is best derived using a falling weight deflectometer survey on the existing rigid pavement.

Falling weight deflectometer (FWD)

The FWD applies to the surface of a pavement a load whose nature bears a close resemblance to that which would be imposed by a travelling vehicle. A series of geophones are located at the point of application of the load and at set distances from it. The purpose of these is to measure the deformations along the surface of the pavement slab caused by the load. The FWD generates a load pulse by dropping a mass onto a spring system, with the weight and drop height adjusted to give the required impact. Peak vertical deflections are measured at the centre of the loading plate and at the several radial positions where the series of geophones are located. On concrete pavements, where deflections may be very low, the load level should be set to a nominal $75\,kN \pm 10\%$ (for flexible/composite pavements, the level should be set at $50\,kN \pm 10\%$).

The falling weight deflectometer is used to assess the structural condition of a highway pavement. It allows the deflected shape of the pavement surface to be measured. Estimates of layer stiffness can be derived from information on this deflected shape together with the thickness and make-up of each of the individual strata. One of the primary uses of the FWD is for the assessment of stiffness of the various layers in terms of its elastic modulus and Poisson's ratio. The layer stiffness survey is then used to assess the equivalent surface foundation modulus, used for the design of the concrete overlay.

During the test, there should be no standing water on the surface of the highway, with the load pulse applied through a 300 mm diameter plate. At least three drops plus a small initial drop for settling the load plate should be made at each test point. Normally the loading plate should be located within the nearside wheelpath of the left-hand lane in order to assess the line of greatest deterioration. Measurements for rigid slabs should be taken at mid-slab locations. The temperature of the pavement should be taken at a depth of 100 mm using an electronic thermometer.

A diagrammatic representation of the falling weight deflectometer, together with a typical deflection profile (termed a 'deflection bowl'), is shown in Fig. 9.7.

The FWD deflection data is tabulated and plotted to illustrate the change in pavement response along the highway. Different pavement layers influence certain sections of the deflection bowl, as shown in Fig. 9.8:

- The central deflection d_1 indicates the overall pavement performance
- d_1 minus d_4 points to the condition of the bound pavement layers
- d_6 is a sign of the condition of the pavement subgrade.

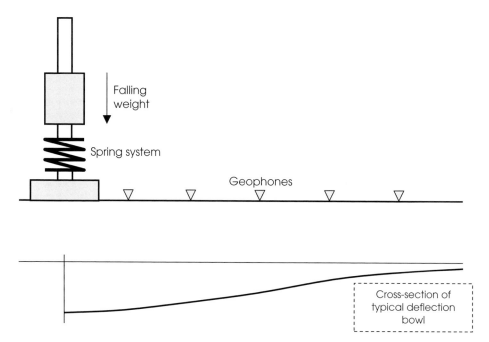

Figure 9.7 Diagrammatic representation of falling weight deflectometer.

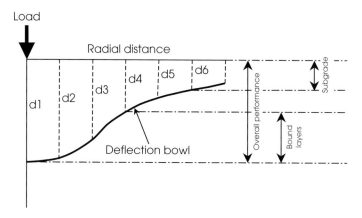

Figure 9.8 FWD deflection profiles and three major indicators d_1, $d_1 - d_4$ and d_6.

The shape of the deflection bowl depends on the type, thickness and condition of the constituent layers within the pavement. A mathematical analysis is then used to match layer stiffnesses to the deflections obtained.

For foundation layers of both concrete and flexible pavements, a layer stiffness of at least 100 MPa is generally associated with good performance of fully flexible pavements and is also thought to be a reasonable value for the unbound foundation layers of both flexible composite and full concrete pavements. These are used directly in the design of a concrete overlay on a rigid pavement. (In the case of a concrete pavement, stiffness will decrease with the proximity of a joint, with cracking, with debonding and as a result of poor compaction.)

Concrete overlays to existing pavements

The quality of the existing surface material and the design of the concrete overlay may require that certain preparatory work be carried out prior to the actual overlaying process.

The level of acceptability of various existing pavement and overlay options together with any required surface treatments are detailed in Table 9.3.

		Existing pavement			
		Flexible/ flexible composite	URC and JRC	CRCP	CRCR
Overlay	URC	2	3	1	2
	JRC	2	3	1	2
	CRCP	1	4	1	1
	CRCR	1	2	1	1

Table 9.3 Acceptability of certain overlay options

JRC: Jointed reinforced concrete
URC: Unreinforced concrete
CRCP: Continuously reinforced concrete pavement
CRCR: Continuously reinforced concrete roadbase
Score 1:   Acceptable with no surface treatment other than remedial works normally necessary
Score 2:   Separation membrane required
Score 3:   No surface treatment other than remedial works is normally necessary, but joints should occur above one another
Score 4:   This combination is not appropriate under normal circumstances
(HD 30/99)

Concrete overlay design

Having determined information on the structural condition of the road via use of the equivalent surface foundation modulus (ESFM), established from an FWD survey, the required overlay thickness can be finalised. It is recommended in HD 30/99 that a representative value for the ESFM is obtained for each section of the highway being considered for treatment, with values taken both along and across the carriageway under examination. In general, the 15th percentile modulus value should be employed for each length of highway undergoing treatment, i.e. the value exceeded by 85% of the sample values established.

Figures 9.9 to 9.12 detail the design thickness for each type of rigid pavement slab. In each case, the required pavement life in millions of standard axles is read off against the ESFM of the existing slab.

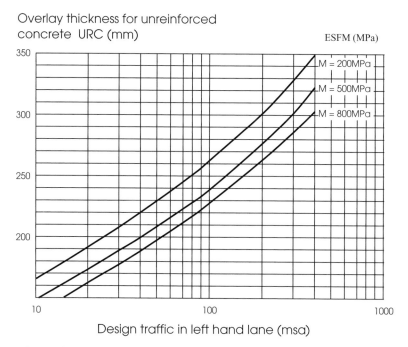

Figure 9.9 Design thickness for unreinforced concrete (URC) overlay (source: HD 30/99) (DoT, 1999).

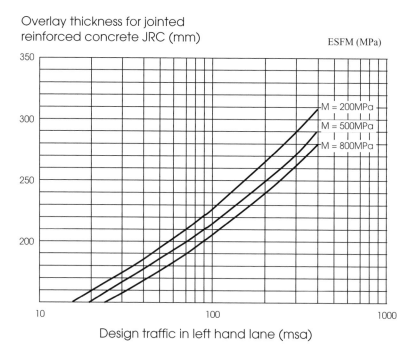

Figure 9.10 Design thickness for jointed reinforced concrete (JRC) overlay (source: HD 30/99) (DoT, 1999).

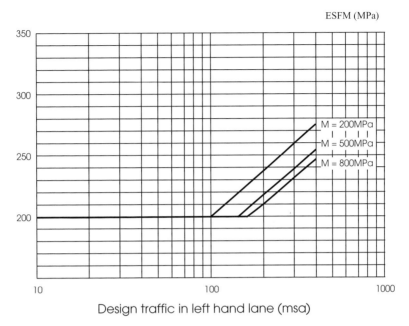

Figure 9.11 Design thickness for continuously reinforced concrete pavement (CRCP) overlay (source: HD 30/99) (DoT, 1999).

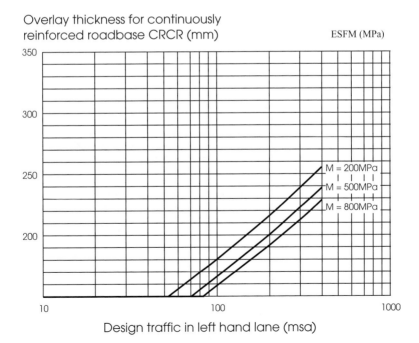

Figure 9.12 Design thickness for continuously reinforced concrete roadbase (CRCR) overlay (source: HD 30/99) (DoT, 1999).

In order both to ensure adequate cover for reinforcement and for practical considerations, the minimum thicknesses of concrete slabs are (HD 26/01) (DoT, 2001):

- 150 mm for unreinforced concrete (URC), jointed reinforced concrete (JRC) and continuously reinforced concrete roadbase (CRCR) slabs
- 200 mm for continuously reinforced concrete pavement (CRCP) slabs

---

Example 9.3 – Design of concrete overlay on rigid pavement
An existing continuously reinforced concrete pavement (CRCP) is required to be overlaid in order to deliver a total pavement life of 200 million standard axles.

Assuming an equivalent surface foundation modulus (ESFM) of 200 MPa, detail two options for overlaying this pavement.

Solution

From Table 9.3, all four types of overlay (URC, JRC, CRCP, CRCR) are acceptable for use over an existing continuously reinforced concrete pavement, with no additional surface treatment necessary. Two options are addressed here:

(1) Jointed reinforced concrete overlay, from Fig. 9.10, for msa = 200 and ESFM = 200, overlay thickness = 270 mm
(2) Continuously reinforced concrete overlay, from Fig. 9.11, for msa = 200 and ESFM = 200, overlay thickness = 220 mm.

9.7   Sideway force coefficient routine investigation machine (SCRIM)

9.7.1   Wet skidding

Skidding resistance is of particular interest in wet conditions when the risk of road accidents is greatest. The skidding resistance of a highway is decreased when the surface becomes wet and a lubricating film of water forms between it and the vehicle tyres. As the thickness of the film increases, the ability of the water to be expelled is decreased. The problem becomes greater as vehicle speeds increase and surface–tyre contact times decrease.

The more effectively the film of water between the tyre and surface can be removed, the greater will be the resistance of the vehicle to skidding. Maintaining adequate tyre treads (a minimum of 1.6 mm is recommended in the UK) is a particularly effective mechanism for removing surface water from a wet highway pavement.

Possession of adequate surface texture is also a valuable preventative device.

The texture depth or 'macrotexture' of the surface refers to the general profile of, and gaps in between, the channels/grooves in the road surfacing. It contributes to skidding resistance, primarily at high speeds, both by providing drainage paths that allow the water to be removed from the tyre/road interface and by the presence of projections that contribute to hysteresis losses in the tyre (this relates to a tyre's capability to deform in shape around the particles of the aggregate within the surfacing, causing a consequent loss of energy). The sand patch test is the oldest method for measuring texture depth. It involves using a known volume of sand to fill the voids in the pavement surface up to their peaks, measuring the surface area covered by the sand and calculating the texture depth by dividing the volume of sand by the area of the patch.

The microscopic texture of the aggregate in the surfacing material is also crucial to skidding resistance. Termed 'microtexture', it relates to the aggregate's physical properties and is important to low-speed skidding resistance. More rounded, smooth particles offer less resistance to skidding than rougher constituents. The action of traffic will tend gradually to reduce a particle's microtexture. Its susceptibility to this wearing action is measured by the parameter *polished stone value* (PSV). First introduced in the UK in the 1950s, the PSV test is the only standard laboratory method of microtexture measurement within the UK (Roe & Hartshorne, 1998). It is a measure of the long-term frictional property of the microtexture.

The test is performed in two parts. First, cubic-shaped, slightly curved, 10 mm test specimens are placed in an accelerated polishing machine, where they are subject to polishing/abrasion for 6 hours. Second, a pendulum friction tester measures the degree to which the specimens have been polished. The result generally lies in the 30 to 80 range, with higher values indicating higher resistance to polishing.

Chippings utilised within a hot rolled asphalt wearing course or as part of a surface dressing process must possess a minimum PSV valuation, depending on the type of site (approach to traffic signal, roundabout, link road) and the daily traffic flows. Sensitive junction locations usually require surfacing materials to have a PSV in the 65 to 75 range, with a general minimum value of 45 required at non-critical locations.

9.7.2   Using SCRIM

As noted earlier in the chapter, SCRIM constitutes a major form of routine assessment of a highway's condition. It measures the skidding resistance of the highway surface that is being gradually reduced by the polishing action of the vehicular traffic (DoT, 1994a).

A vehicle will skid whenever the available friction between the road surface and its tyres is not sufficient to meet the demands of the vehicle's driver. SCRIM was pioneered in the 1970s to provide a methodology for measuring the wet skidding resistance of a highway network.

For wet surfaces, the sideway force coefficient (SFC) is speed dependent. The equipment is capable of testing between 20 and 100 km/hr; the target testing speed is 50 km/hr. The permitted range for testing is 30 to 67 km/hr with a standard speed correction applied if the vehicle deviates from the 50 km/hr target. The SFC evolved from the motorcycle-based testing machines in the 1930s when it was found that the force exerted on a wheel angled to the direction of travel and maintained in this vertical plane with the tyre in contact with the surface of the highway, was capable of correlation with the resistance to wet skidding of the pavement surface. The sideway force derived in this manner is defined as the force at 90° to the plane of the inclined wheel. It is expressed as a fraction of the vertical force acting on the wheel. The SCRIM apparatus consists of a lorry with a water tank and a test tyre, made of solid rubber, inclined at an angle of 20° to the direction of travel and mounted on an inside wheel track. Water is sprayed in front of the tyre in order to provide a film thickness of constant depth. The sideway force coefficient is obtained by expressing the measured sideway force exerted on the test wheel as a fraction of the vertical force between the test wheel and the highway.

A diagrammatic representation of the SCRIM apparatus is shown in Fig. 9.13.

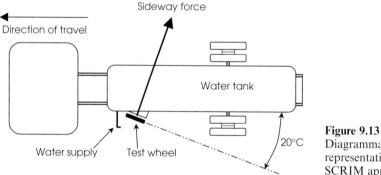

Figure 9.13 Diagrammatic representation of SCRIM apparatus.

The SFC of a highway pavement depends on traffic flow, seasonal variations and temperature. SCRIM coefficients have a typical range of 0.30 to 0.60. A typical mean summer value would be 0.5.

9.7.3   Grip tester

A grip tester is widely used on public roads. It can be pushed by hand or towed behind a vehicle that can travel at speeds of up to 130 km/hr. A measuring wheel rotates at a slower rate than the main wheels of the apparatus and a film of water is sprayed in front of the tyre. A computer lodged within the apparatus monitors both the vertical load and the frictional drag acting on the measuring wheel.

The grip tester is easily operated and maintained, with lower running costs than the SCRIM apparatus. Its ability to be hand-pushed enables it to be used in pedestrian areas.

A diagrammatic representation of the grip tester is shown in Fig. 9.14.

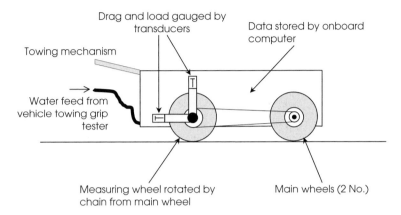

Figure 9.14 The grip tester.

The results obtained from the grip tester are consistent with those from SCRIM and a strong correlation between the two tests has been established.

9.8   References

Addis, R.R. & Robinson, R.G. (1983) Estimation of standard axles for highway maintenance. *Proceedings of Symposium on Highway Maintenance and Data Collection*, University of Nottingham, UK.

Asphalt Institute (1983) *Asphalt Overlays for Highway and Street Rehabilitation*. Manual Series No. 17 (MS-17). The Asphalt Institute, Maryland, USA.

Boussinesq, J. (1885) *Application of Potentials to the Study of Equilibrium and Movement of Elastic Solids*. Gauthier-Villars, Paris.

Claessen, A.I.M. & Ditmarsch, R. (1977) Pavement evaluation and overlay design – The Shell Method. *Proceedings of the 4th Conference on the Structural Design of Asphalt Pavements*, Volume 1. Ann Arbor, University of Michigan, USA.

DoT (1994a) Skidding resistance, HD 28/94. *Design Manual for Roads and Bridges, Volume 7: Pavement Design and Maintenance*. The Stationery Office, London, UK.

DoT (1994b) Structural assessment methods, HD 29/94. *Design Manual for Roads and Bridges, Volume 7: Pavement Design and Maintenance*. The Stationery Office, London, UK.

DoT (1994c) Maintenance of bituminous roads, HD 31/94. *Design Manual for Roads and Bridges, Volume 7: Pavement Design and Maintenance*. The Stationery Office, London, UK.

DoT (1994d) Maintenance of concrete roads, HD 32/94. *Design Manual for Roads and Bridges, Volume 7: Pavement Design and Maintenance*. The Stationery Office, London, UK.

DoT (1999) Maintenance assessment procedure, HD 30/99. *Design Manual for Roads and Bridges, Volume 7: Pavement Design and Maintenance*. The Stationery Office, London, UK.

DoT (2001) Pavement design and construction, HD 26/01. *Design Manual for Roads and Bridges, Volume 7: Pavement Design and Maintenance*. The Stationery Office, London, UK.

Kennedy, C.K., Fevre, P. & Clarke, C. (1978a) *Pavement Deflection: Equipment for Measurement in the United Kingdom*. TRRL Report LR 834. Transport and Road Research Laboratory, Crowthorne, UK.

Kennedy, C.K. & Lister, N.W. (1978b) *Prediction of Pavement Performance and the Design of Overlays*. TRRL Report LR 833. Transport and Road Research Laboratory, Crowthorne, UK.

Mayhew, H.C. & Harding, H.M. (1987) *Thickness design of concrete roads*. Department of the Environment, Department of Transport, TRRL Report RR 87. Transport and Road Research Laboratory, Crowthorne, UK.

Roe, P.G. & Hartshorne, S.E. (1998) *The polished stone value of aggregates and in-service skidding resistance*. TRL Report 322, Transport Research Laboratory, Crowthorne, UK.

Index